iPHONE FOR NON-TECH- SAVVY SENIORS

TABLE OF CONTENTS

DISCLAIMER NOTICE:

Please note the information contained within this document is for educational and entertainment purposes only. All effort has been executed to present accurate, up to date, reliable, complete information. No warranties of any kind are declared or implied. Readers acknowledge that the author is not engaged in the rendering of legal, financial, medical or professional advice. The content within this book has been derived from various sources. Please consult a licensed professional before

attempting any techniques outlined in this book. By reading this document, the reader agrees that under no circumstances is the author responsible for any losses, direct or indirect, that are incurred as a result of the use of the information contained within this document, including, but not limited to, errors, omissions, or inaccuracies.

We invite you to scan this **QR code** using the camera of your phone to access your bonus content:

SCAN THE QR CODE BELOW

You will access to **2 EBOOKS:**

1. **"Health Tips for Seniors":** The top 10 essential life tips for seniors! These strat-egies will help you look and feel younger in a matter of a few weeks!
2. **"Aging Secrets":** The Secrets To Living Longer By Looking Younger And Feel-ing Younger (Backed by science)

INTRODUCTION

iPhone is a series of smartphones made by the tech giant Apple and is popular for its simple and user-friendly design. Therefore, it is a great choice for senior citizens who use smartphones. While this device is easy-to-use overall, it can also lead to common issues that older people have while using technology.

Most seniors need help with smartphones due to their advanced age and possibly limited thinking processes. Mastering new technologies are sometimes tricky because elders lack experience with technology. Seniors often have a lesser basis from which to absorb new information. They often feel uneasy while tapping the touch screens for fear of something unpleasant happening. Menus and user interfaces may be complicated and login credentials like passwords and usernames might be challenging to remember.

Because of this, the elderly fear being regarded as uneducated by others, and as a result, they become disinterested in using the iPhone and exploring its services.

If you're a senior, and you've been frustrated because your iPhone experience didn't work the way you wanted it to, don't worry; there's an easy way to set up your iPhone and ensure it works for you.

iPhone for Non-Tech-Savvy Seniors is the most complete and easy-to-understand guide to learning how to use your iPhone SE, 13, 14, or 15. Make the most of it by following this book's step-by-step instructions, helpful tips, and guidance to get you started.

Whether you are a senior just learning about the iPhone for the first time or a seasoned user looking to take it to the next level, this book will teach you how to master your iPhone quickly and easily.

Learn how to use the most exciting features of the phone, like Siri and Face ID, and how to manage iPhone media and send messages. This book includes privacy and security information to ensure that personal information is safe. You will also learn some handy tips on how to use iPhone apps and how to fix the most common problems with iPhones.

This book was created to help you learn the ins and outs of your iPhone and to fit your needs in using technology so you can have the most fun and meaningful life. We believe technology should be easy and intuitive, so we've made it our mission to help seniors set up iPhones simply and securely.

Boost your confidence with your iPhone by reading this book. No more worries about using your device and have a happy and healthy retirement. Control your phone with your thumbs until you learn how to take group photos. Stay connected; if an emergency happens, you can call and chat with friends and family to stay safe.

If this might be what you are looking for, get your hands on the guide now!

CHAPTER 01

MEET YOUR IPHONE

Apple releases its latest version of iPhone products annually to provide the best smartphones it can develop The brand ensures the specifications of its newest iPhone include high-quality internet connectivity, longe battery life, improved camera, etc., than the previously released versions. This upgrade in the annua product cycle of iPhones is known as a generation.

Since its kickstart in 2007, Apple has brought 16 generations of iPhones, having its phone models launche in order. In 2021, the iPhone 13 lineup was released, followed by iPhone SE and 14 series in 2021. Eac smartphone lineup has four varieties: the basic iPhone, Plus, Pro, and Pro Max. As the name suggests, th baseline model among the four is the most affordable, while the latter is the most expensive.

Although another iPhone is launched every year, each model among the product lines has heavy similaritie but still has considerable differences. In the next topic, you will understand the main similarities an differences between the iPhones SE, 13, and 14.

Comparing iPhone SE, 13 and 14

Apart from the prices, the main thing users check out about iPhone is the upgrade and other specificatio between the handsets. iPhones SE, 13, and 14 have significant similarities and notable differences in thes models. Knowing the basic specifications of your iPhone is essential to help you maximize your phon experience.

Screen

iPhones 13 and 14 have the same screen size of 6.1 inches, unlike the iPhone SE, which is only 4.7 inches. The screen display of both iPhones 13 and 14 is also the same in sharpness, which is better than that of the iPhone SE. If you wish to carry a lighter smartphone, then the 144g iPhone SE is the one for you. iPhones 13 and 14 are models larger in size and bulky, weighing 173g and 174g, respectively.

Storage

Regarding storage, there are three options for the iPhone SE. The option with the least space would be best for someone who uses their phone occasionally. However, if you want to take and save as many photos and videos as you want without worrying about running out of space, you can choose the option with the most space. On the other hand, the storage options for the iPhones 13 and 14 are the same. The most space either model can hold is almost twice as much as the iPhone SE. This means you will have plenty of room to download and save all media on your iPhone, like videos, eBooks, music, applications, and games.

Remember that less storage space slows down your iPhone, so the more memory, the better.

Sensor

The iPhone 13 and 14 use a Face ID to recognize or verify their user's identity. It doesn't have a Touch ID sensor like iPhone SE because the latter has a Home button at the bottom of its screen. The later chapter will explain setting up Face ID and Touch ID.

Battery life

One of the most important things to know is how long your device's battery will last. The latest iPhones have impressive battery performance. iPhone 14 can play a downloaded video for up to 20 hours and up to 16 hours of streaming video over the internet, while iPhone 13 manages to play a local video for up to 19 hours and up to 15 hours of streaming video over the internet, which is just slightly less than the former. Finally, iPhone SE can play a downloaded video for up to 15 hours and up to 10 hours of streaming video online.

There are several ways you can charge these smartphones. Connect a compatible power adapter, then plug a Lightning cable into your phone. If you prefer wireless charging, you can use Qi, which allows you to charge your device from a distance without a cable. Another wireless charging option is Apple's MagSafe which can quickly and efficiently charge your phone. The iPhone SE supports Qi, while the iPhone 13 and 14 support Qi and MagSafe charging.

Performance and other details

The speed and efficiency are high, as expected of iPhone models, so you will enjoy smooth menu navigation on your phone. The overall performance between iPhones 13, 14, and SE is roughly the same.

The sim card slots available for iPhone 14 are one physical and two digital sims. Both iPhone 13 and SE support one physical sim and one digital sim. Unfortunately, Apple removed the headphone port in 2016 so you need to connect a wireless headphone or earphone to your iPhone if you don't want to listen via a speaker.

CHAPTER 02

SETTING UP YOUR iPHONE

After your iPhone purchase, you next need to set up your device to use it according to your preference. This chapter will walk you through an easy way of setting up your iPhone for everyday use. Many of Apple's product lines have the same setup instructions that fit iPhones 13, 14, and SE.

Smartphone users love to personalize their phone's access — and with iPhone, you can customize preferences in the display, accessibility, and navigation so you can enjoy more of its services. For a smooth setup of your device, make sure to prepare the following:

- A Wi-Fi connection or mobile data through a carrier
- Any of your old iPhones or old Android phones (for data transfer)
- Old and new device chargers in case of low power

Tip: To use VoiceOver during setup, triple-click the **side button** (iPhones 13 and 14) or the **Home button** (for iPhone SE). To use Zoom, double tap the screen with three fingers to activate.

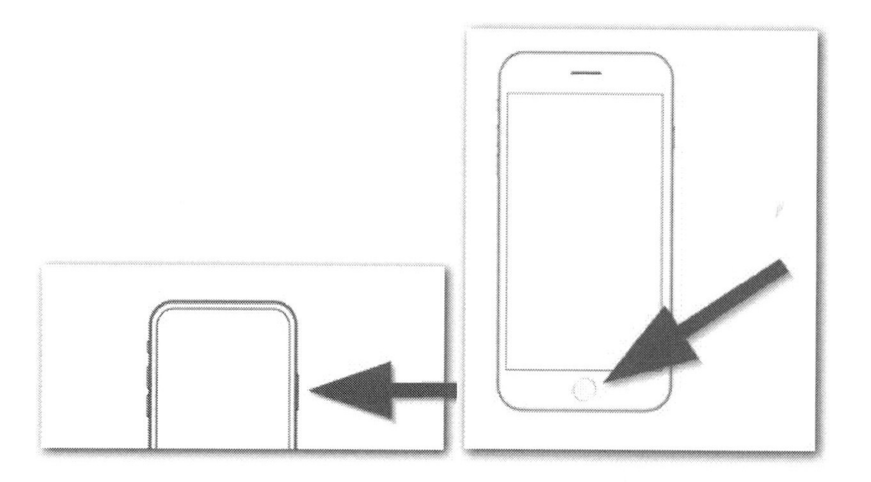

1. Long press the **side button** and wait for the Apple logo to show.

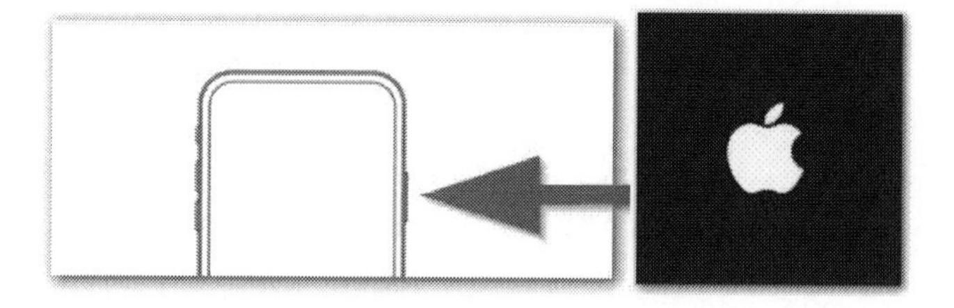

2. "Hello" appears in several languages. Swipe up to continue, then select a language for your device.

3. Select your country or region. You are taken to the Quick Start screen. **Quick Start** enables you t move data directly from your old iPhone to your new one. In this tutorial, select **Set Up Manually** a the bottom of the screen to set up your iPhone and move data manually.

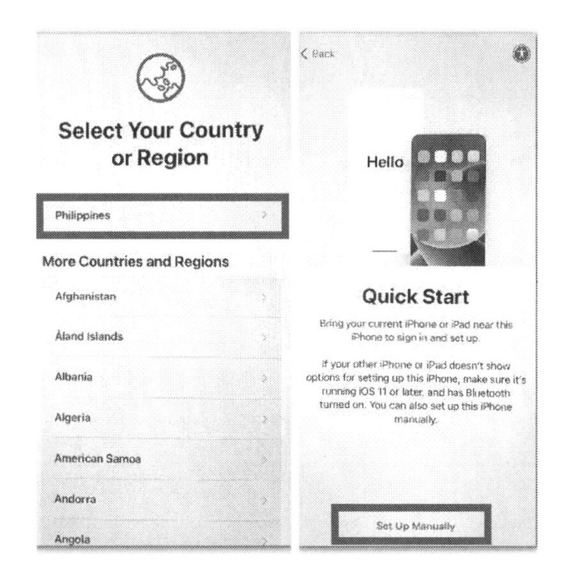

4. On Written and Spoken Languages, tap **Continue** if you want to use the suggested settings on the screen. To customize them individually, tap **Customize settings**, then follow the prompts on the screen.

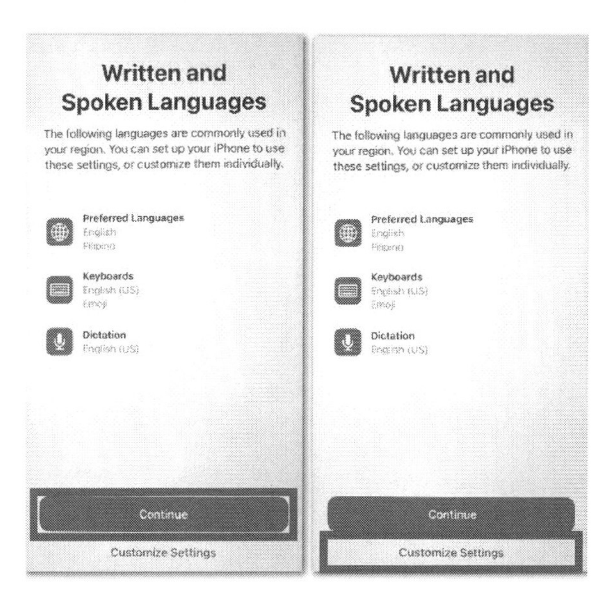

5. Select a **Wi-Fi network** to connect to the internet, then enter the password for the Wi-Fi network selected. Tap **Join**.

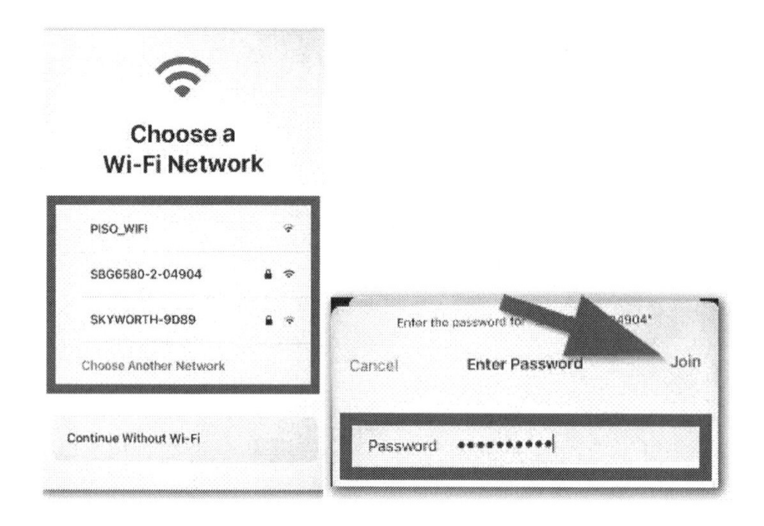

6. Alternatively, you may select **Continue Without Wi-Fi** if you're using cellular data service. Wait fo a few minutes to activate your phone. On the Data & Privacy screen, select **Continue** or **Learn More.**

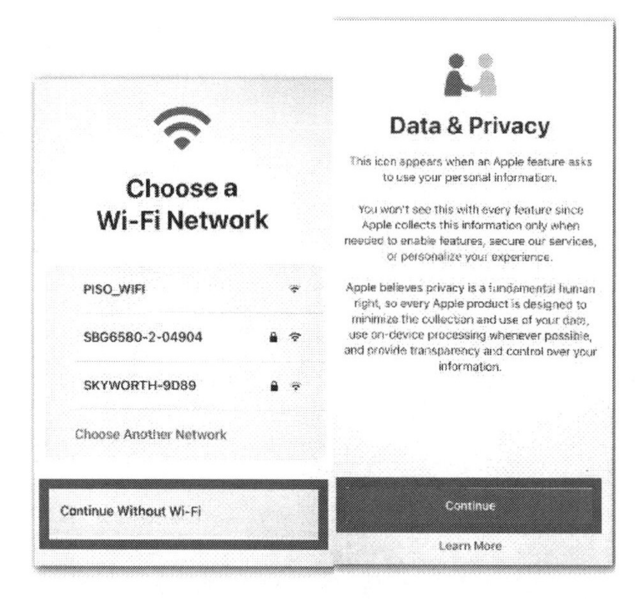

7. Skip the Touch ID (for iPhone SE) or Face ID (for iPhone 13 and 14) setup and select **Set Up Late** You will learn how to set one up later. Skip creating a passcode and select **Passcode Option.**

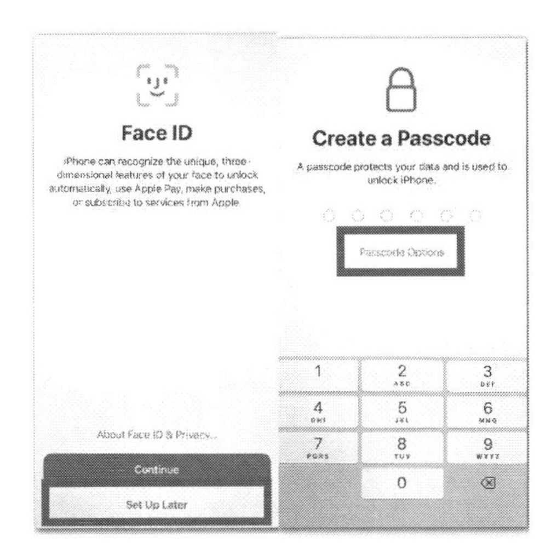

8. Tap **Don't Use Passcode.** You will learn how to set up a passcode later.

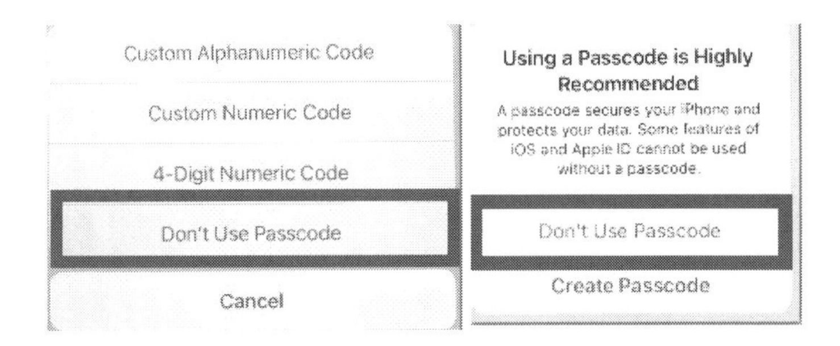

9. Select how you want to restore or transfer data to your new iPhone. Refer to the steps below and select the one you wish to follow. Note: If you don't want to transfer data, tap **Don't Transfer Apps and Data.**

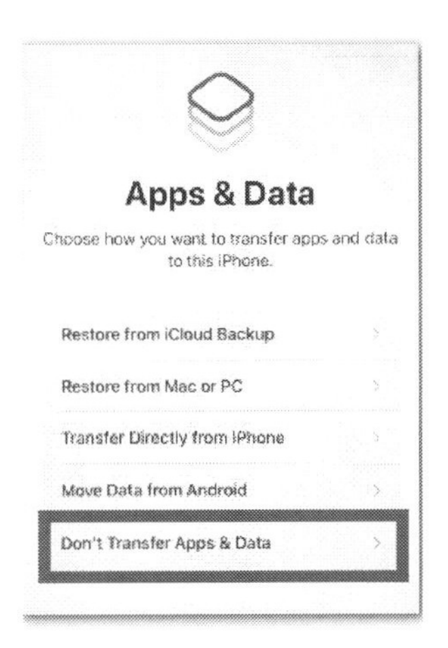

A. **A. Restore from iCloud Backup** – tap if you have backup data in iCloud. iCloud is a data storage software available on every Apple device. If you don't have backup data in iCloud, you can still do it by following these steps:

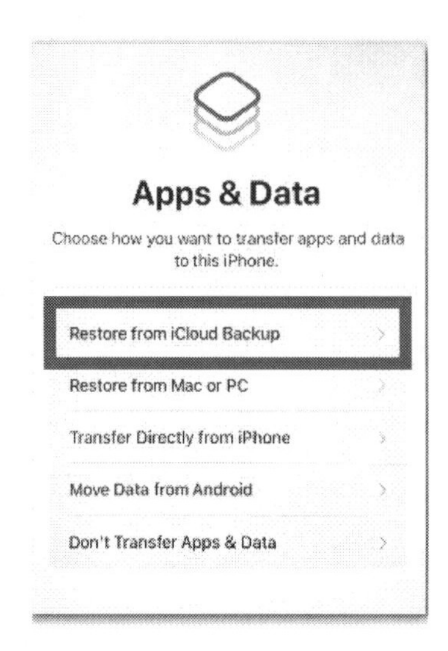

a. Using your old iPhone, Go to **Settings**.

b. Tap **your name** at the top of the screen. Tap **iCloud.**

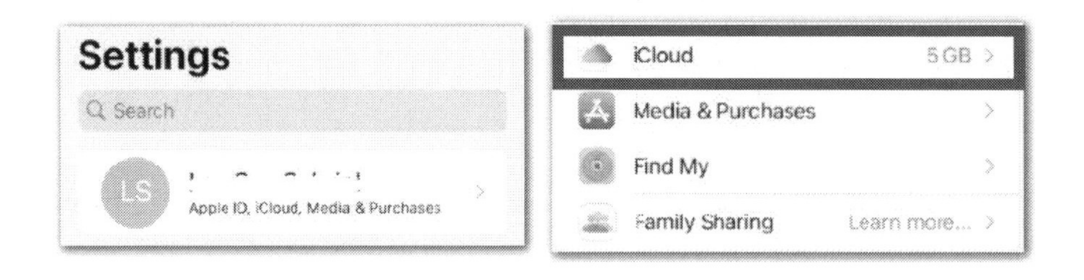

c. Select **iCloud Backup**, then switch on **Back Up This iPhone**.

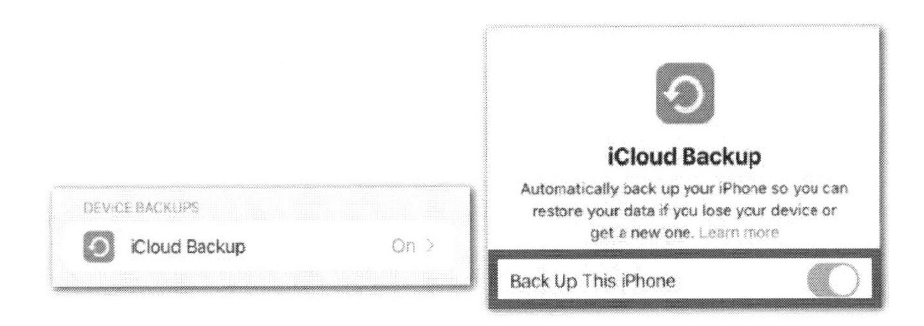

d. Select **Back Up Now**.

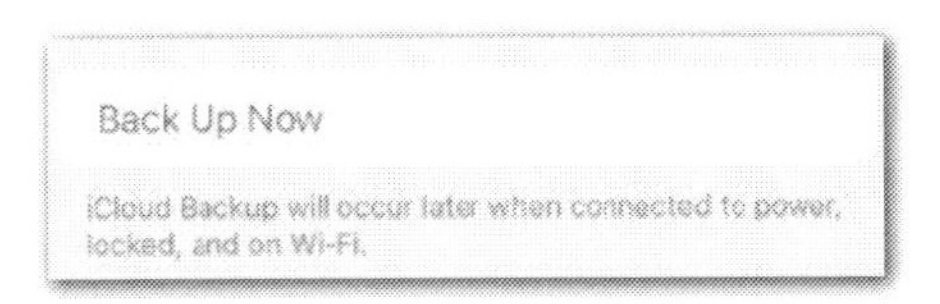

1. Go back to your new device.
2. Sign into iCloud using your Apple ID. Then, tap **Next** at the top-right corner of your screen and enter your Apple ID password.

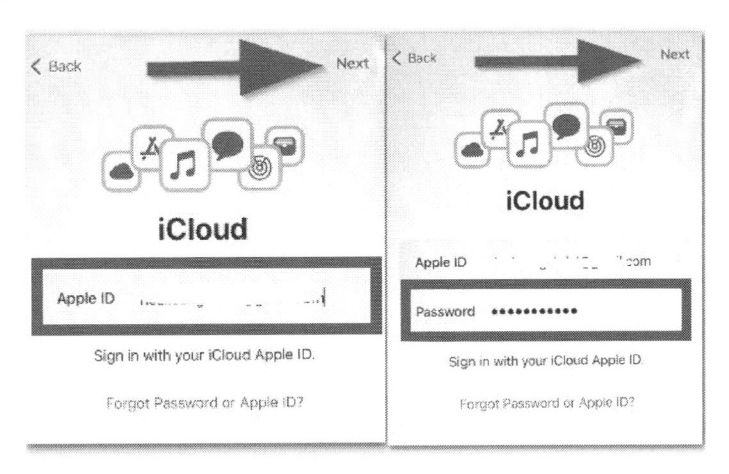

3. If you have forgotten your Apple ID or password, tap **Forgot Password or Apple ID**.

To recover your Apple ID:

a. Tap **Forgot Apple ID.**

b. Enter your information as required on the screen. Tap **Next** at the top-right corne of your screen. After your Apple ID is found, tap **Continue to Sign In.**

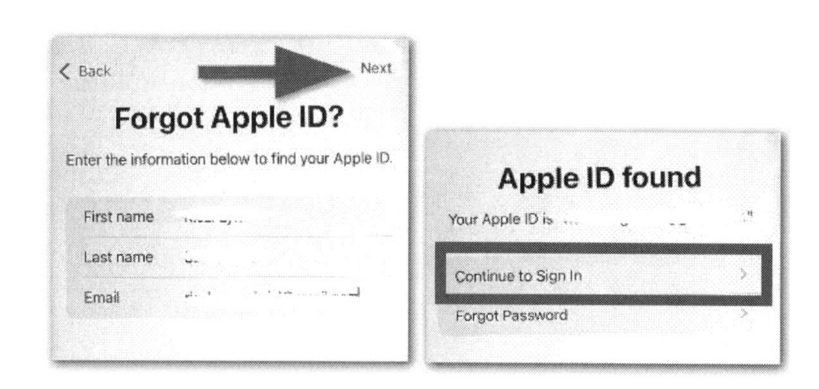

To recover password:

a. Enter your Apple ID, then tap **Next** at the top-right corner of your screen. Confirm your phone number, then tap **Next** at the top-right corner of your screen.

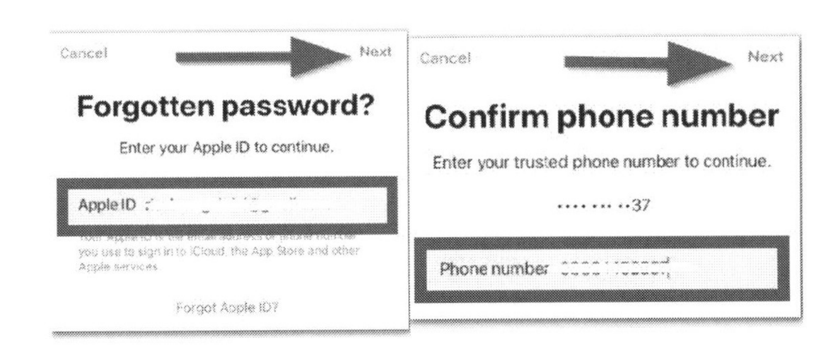

b. Enter the passcode you use to unlock your old iPhone. If you don't know the passcode, tap **Don't know iPhone Passcode?** then follow the prompt on the screen.

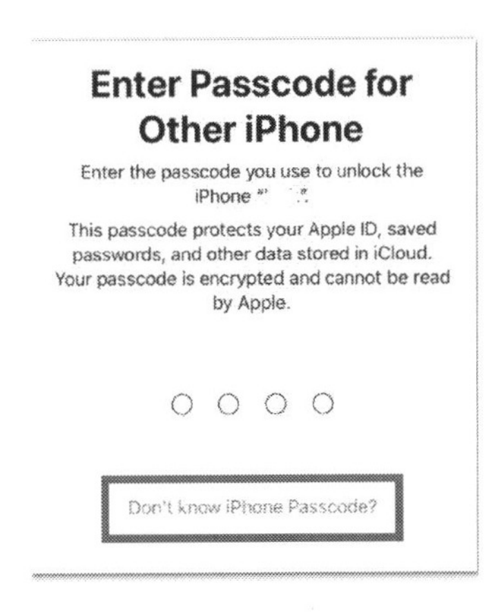

c. Enter the New Apple ID password. Follow the instructions to create a password, then tap **Next** at the top-right corner of your screen.

4. On the Terms and Conditions screen, tap **Agree**. Wait for the next prompt. On Choose Backup screen, select the backup you wish to restore.

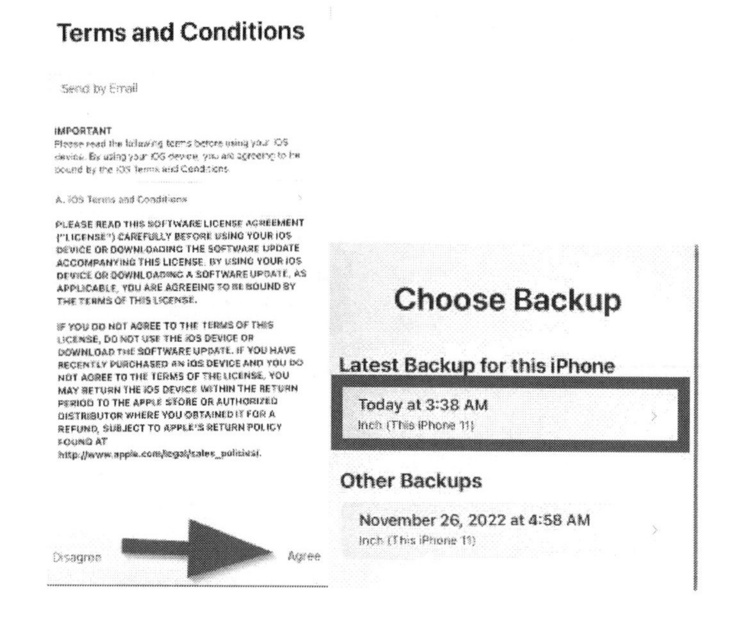

5. On the Make This Your New iPhone screen, review the data included in the restoration. Tap **Continu** to proceed or **Customize settings** if you wish to change these settings.

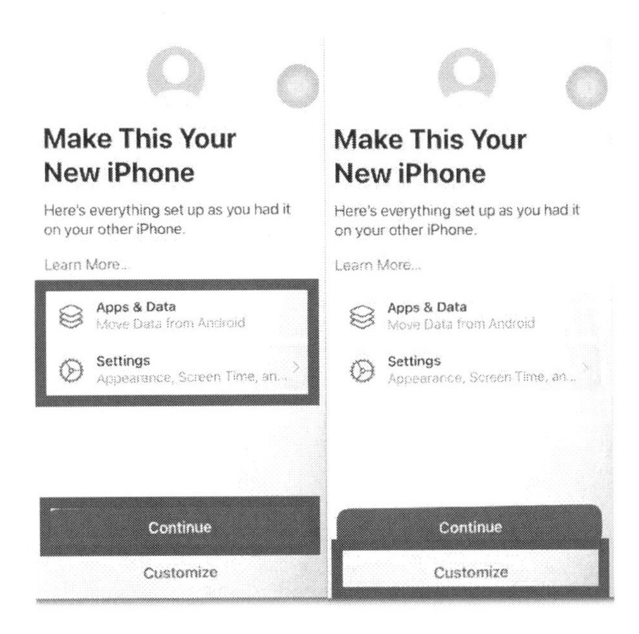

6. Keep your Wi-Fi connection active and watch for the progress bar to reach the end. Disconnecting from Wi-Fi will pause the transfer progress. When the restore is finished, tap **Continue**.

7. Skip the Touch ID (for iPhone SE) or Face ID (for iPhone 13 and 14) setup and select **Set Up Later**. You will learn how to set one up later. On the Terms and Conditions screen, tap **Agree**.

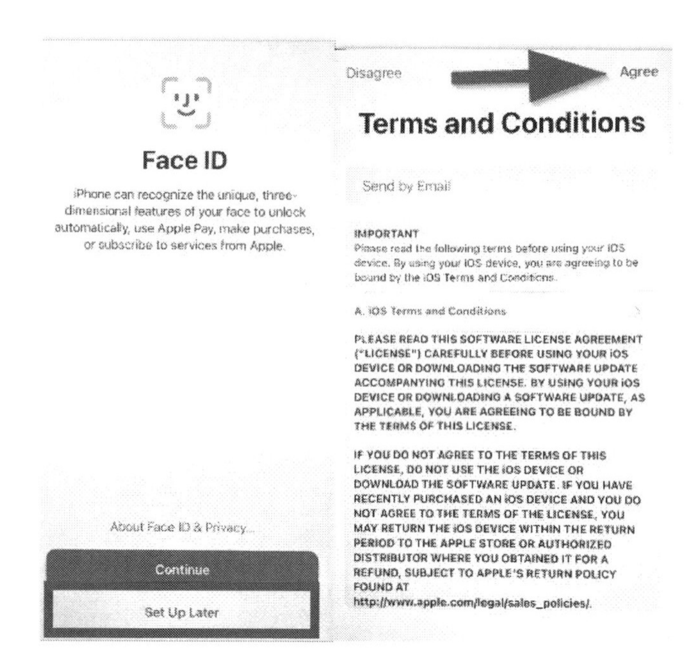

8. On the Improve Siri and Dictation screen, tap **Not Now**. You will learn how to set up and navigate Siri and Dictation later.

9. Your device is ready. Swipe up to get started.

B. **Restore from Mac or PC** – tap if you want to restore from a backup on your computer.

1. If you're using a Mac with macOS Catalina or later versions, click the **Finder**. If you're using a Mac with macOS Mojave, earlier versions, or another computer, click the **iTunes** app.

2. In this tutorial, the device is connected to iTunes. Use a USB cord to connect your phone to your computer. When the prompt appears, click **Continue**.

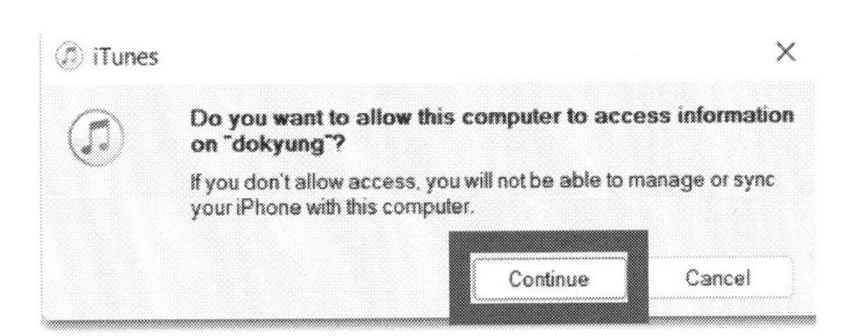

3. On your iPhone, tap **Trust**. If prompted to create a passcode, tap **Don't Use Passcode**. You will learn how to create this later.

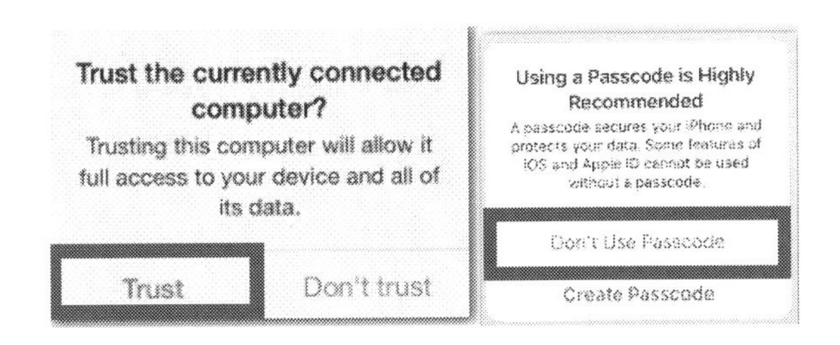

4. When your device appears in iTunes, click the **tiny phone icon** on the top. Select **Restore from this backup**, then **Continue**. If you're using Mac, locate your iPhone in the Finder window, then select the restore option.

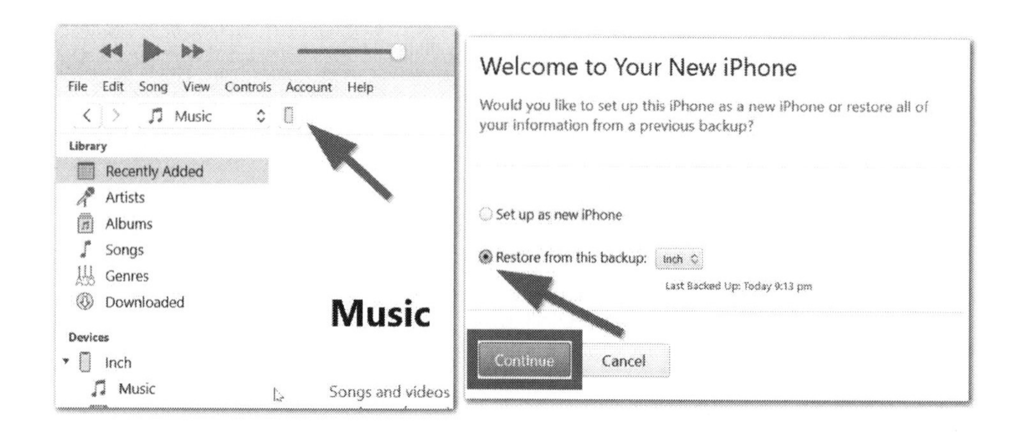

5. Wait for the progress to finish. On your iPhone, tap **Continue** when the restore is completed. Disconnect after the sync is done. Skip the Touch ID (for iPhone SE) or Face ID (for iPhone 13 and 14) setup and select **Set Up Later**. You will learn how to set one up later.

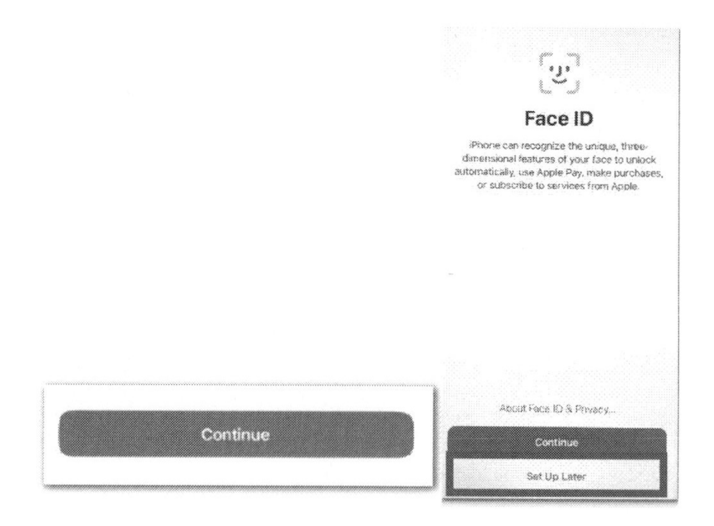

6. Skip creating a passcode and select the **Passcode Option.** Tap **Don't Use Passcode.**

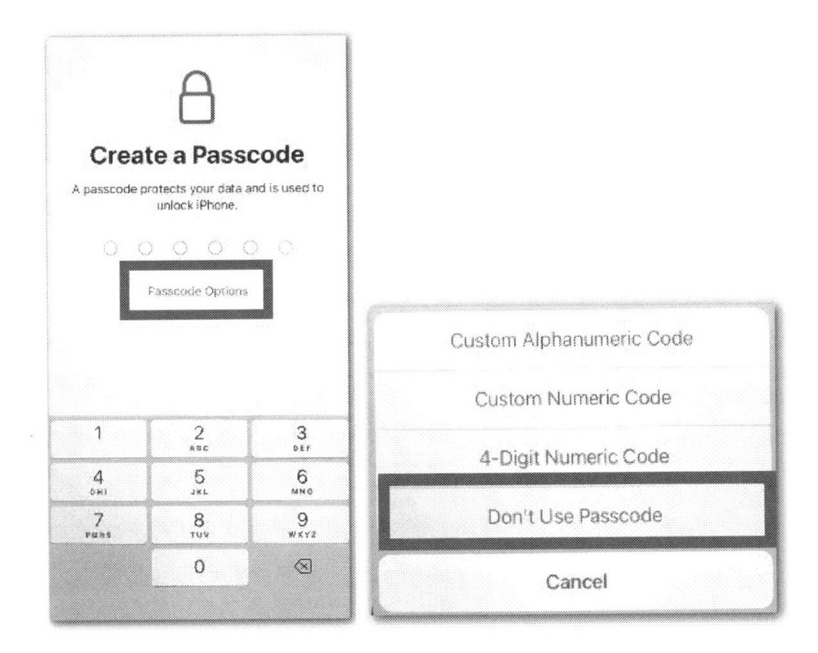

7. You will learn how to set up a passcode later.

8. Enter your Apple ID password, then tap **Sign in**. On the Terms and Conditions screen, tap **Agree**.

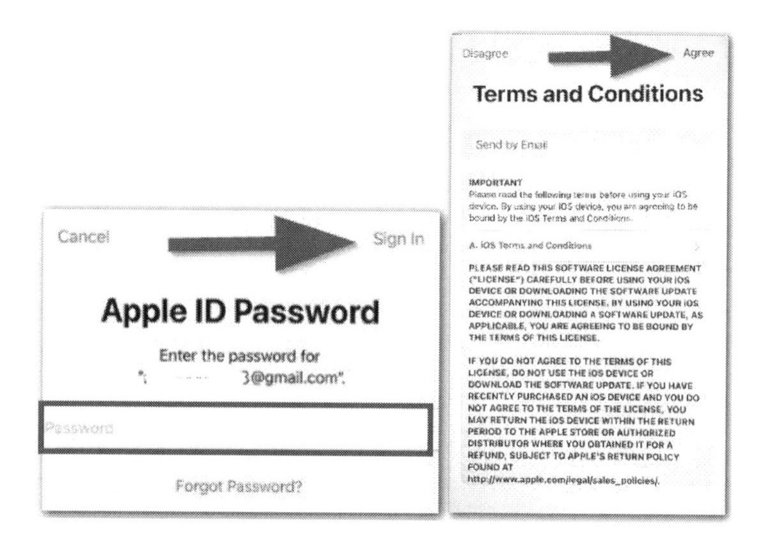

9. On the Location Services screen, select between **Enable location services** or **Disable location services.**

10. Your device is ready. Swipe up to get started.

C. **Transfer Directly from iPhone** – tap if you want to move your data from an old iPhone to your new iPhone. This redirects you to the Quick Start screen.

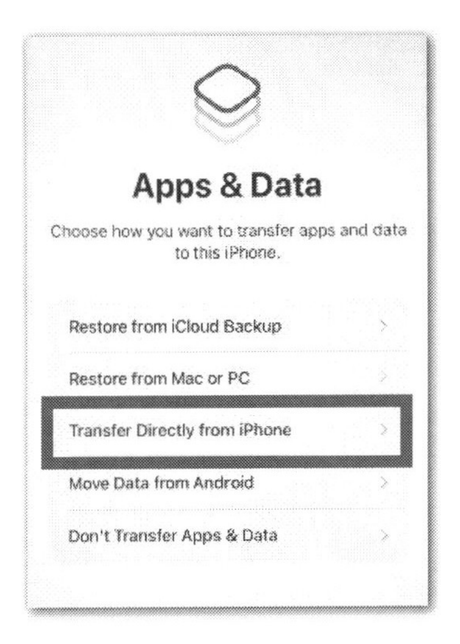

1. Connect your old device to the internet and turn on the **Bluetooth**.

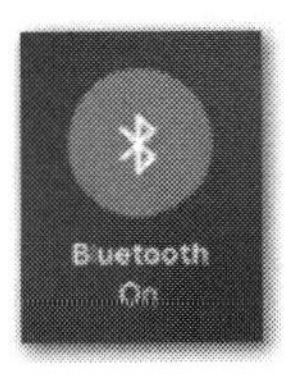

2. Place it near your new device. On your new device, the Apple ID shows from your old phone. Tap **Continue**.

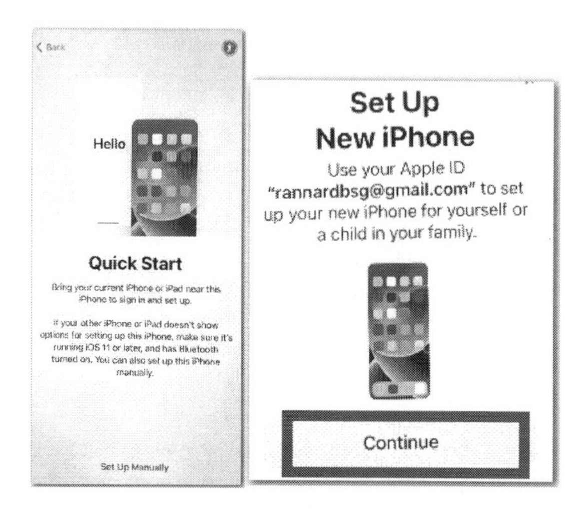

3. Your new phone will display an animation. To proceed, center it in the camera viewfinder of your old device. If the camera isn't available, tap **Authenticate Manually**.

4. On your old device, tap **Set Up for Me.** On your new device, enter the passcode you use to unloc your old iPhone.

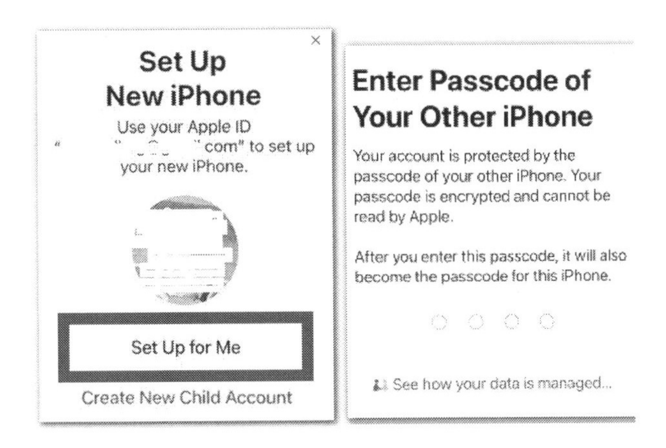

5. Tap **Set Up Later** to skip the setup of Touch ID or Face ID. You will learn how to set one up later.

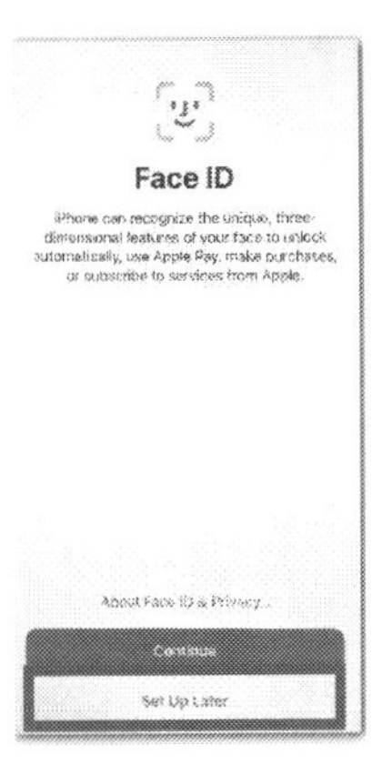

6. Your device is ready. Swipe up to get started.

D. **D. Move Data from Android** – Tap if you want to transfer data from an Android device.

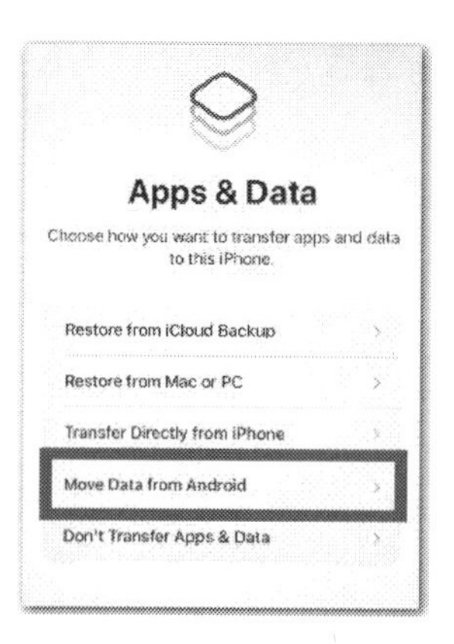

1. Place your iPhone near your Android device.
2. On your Android device, connect to the internet and download a third-party app, such as **Move to iOS** app, that will help move data from your Android device to your new phone. Tap **Move to iOS**. Follow the prompt on the screen.

3. On your iPhone, tap **Continue** on the Move from Android screen. Wait for a code to show.

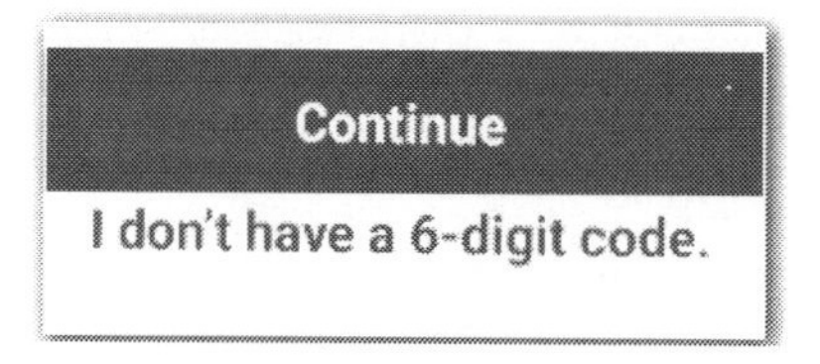

4. Go back to your Android device, then enter the **code**.
5. You will see a temporary network on your iPhone. On your Android device, tap **Connect** to join that network. Wait for the Transfer Data screen to show.

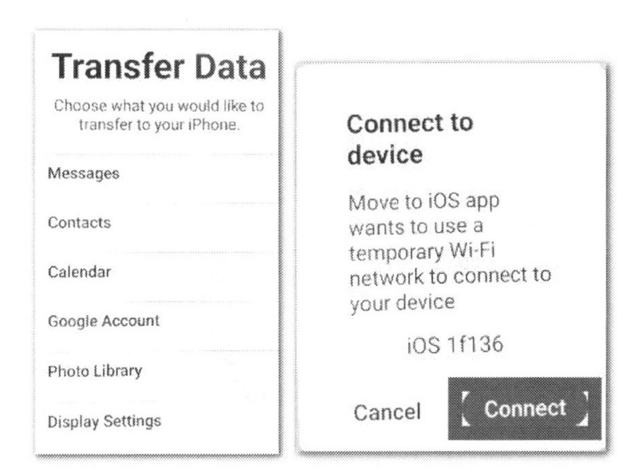

6. Select the data you want to transfer to your iPhone, then tap **Continue**.

7. Wait for the progress bar on your new device to reach the end while keeping your devices close to each other. Expect the transfer to take a while if you move a large content.
8. Make sure that the Move to iOS app remains on the screen to complete the transfer successfully. On your Android device, tap **Done** after the transfer is completed.

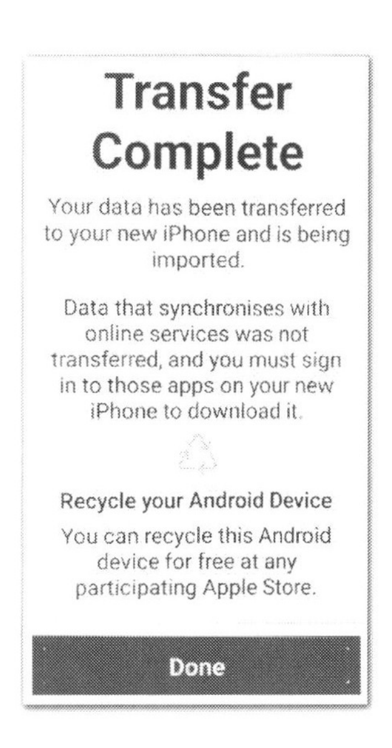

9. On your iPhone, tap **Continue**.

10. Skip the Apple ID setup and tap **Forgot password or don't have an Apple ID?** to set up your Apple ID later. Tap **Don't Use**.

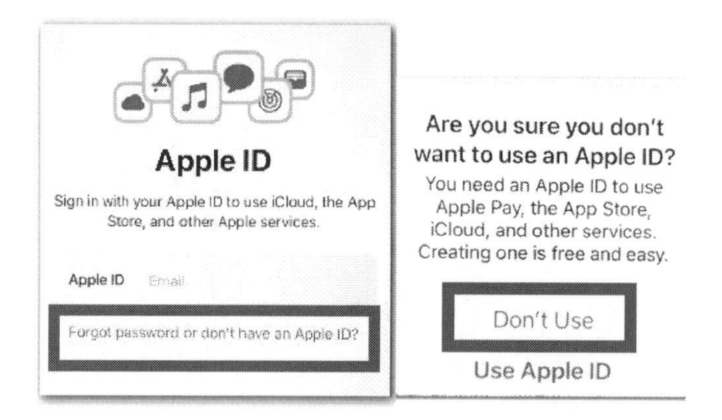

11. On the Terms and Conditions screen, tap **Agree**.

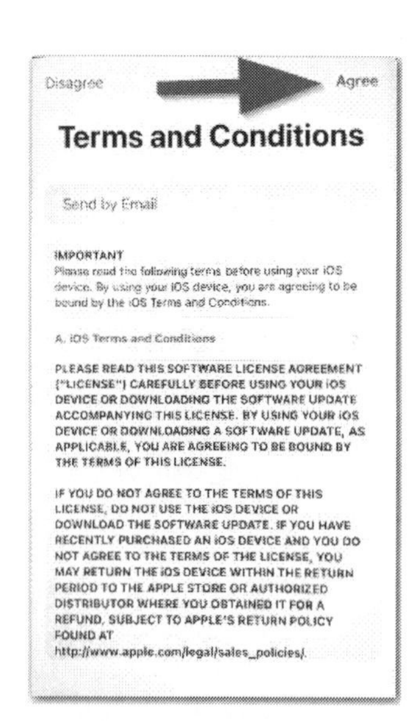

12. On the Make This Your New iPhone screen, review the data included in the restoration. Tap **Continu** to proceed or **Customize settings** if you wish to change these settings.

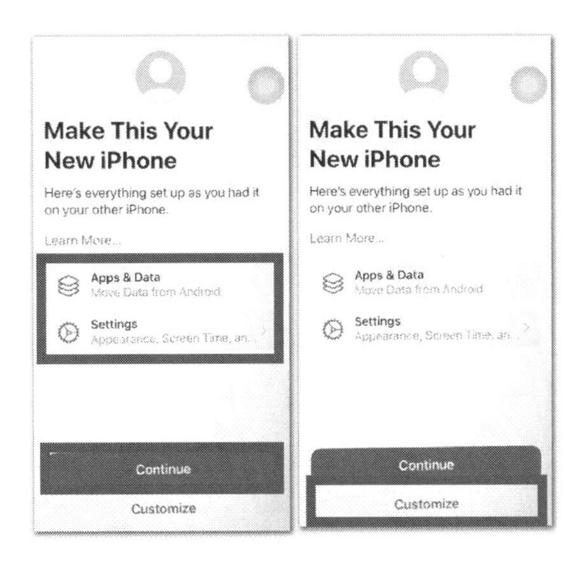

13. On Keep Your iPhone Up to Date, tap **Continue**. Skip iMessage & FaceTime and tap **Continue** to set up later.

14. Your device is ready. Swipe up to get started.

15. Continue these steps if you selected **Don't Transfer Apps and Data**. Otherwise, move to How to Se Up a Passcode. Skip the Apple ID setup and tap **Forgot password or don't have an Apple ID?** Ta **Set up Later in Settings**.

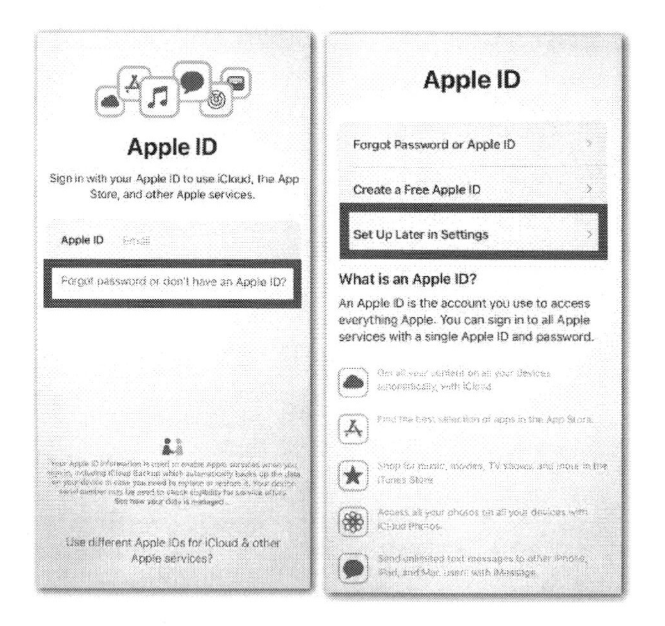

16. On the Terms and Conditions screen, tap **Agree**. On the Keep Your iPhone Up to Date screen, ta **Continue**.

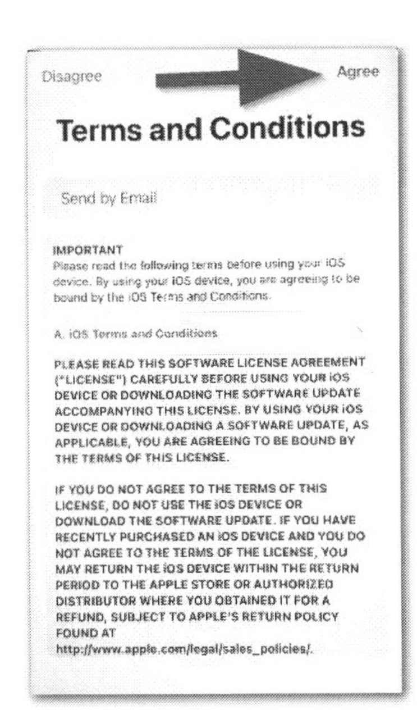

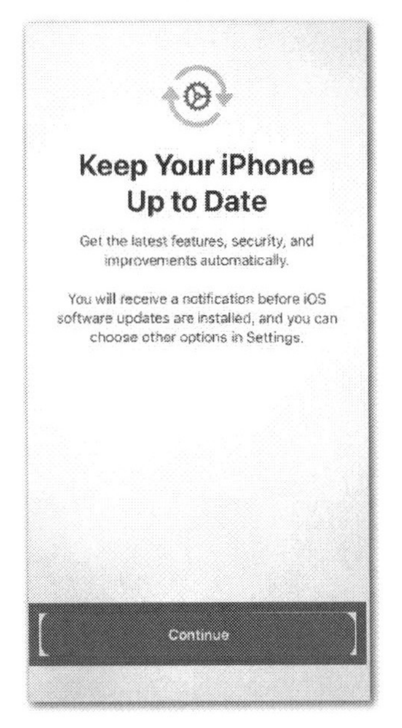

17. On the Location Services screen, select between **Enable location services or Disable location services**. On the Siri screen, tap **Set up Later in Settings**.

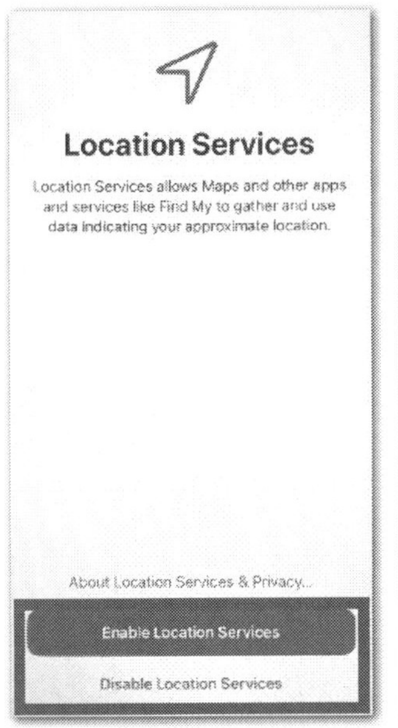

18. On the Screen Time screen, tap **Set Up Later in Settings**. You will learn how to set one up later. On the App Analytics screen, select between **Share with Apple** and **Don't Share**.

Note: Details regarding operating system characteristics, performance statistics, and data about how you use your devices and applications may be included in iPhone Analytics. This data is collected and utilized by Apple to improve and develop its products and services while keeping its users' identification private.

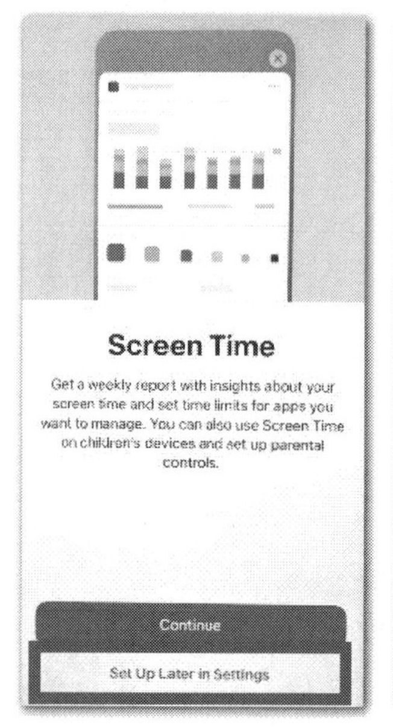

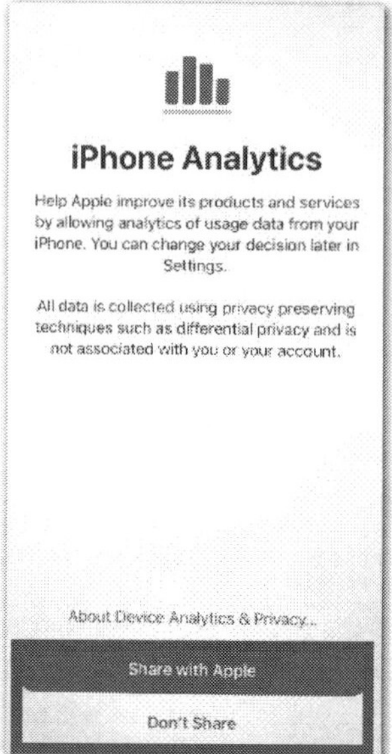

19. On the Appearance screen, select between **Light** or **Dark**. Tap **Continue**.

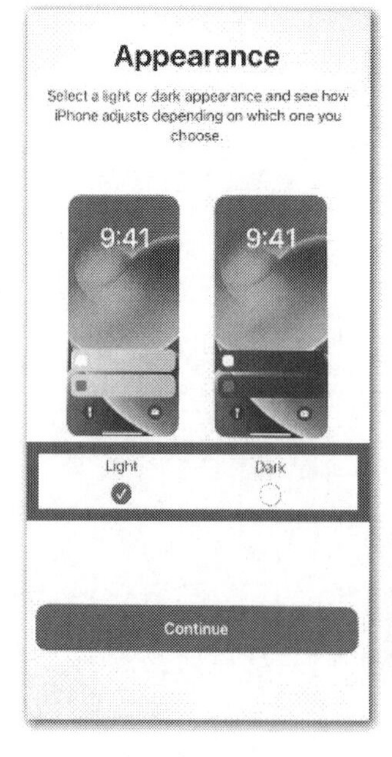

20. On the Display Zoom screen, select between **Standard** or **Zoomed**. Tap **Continue**.

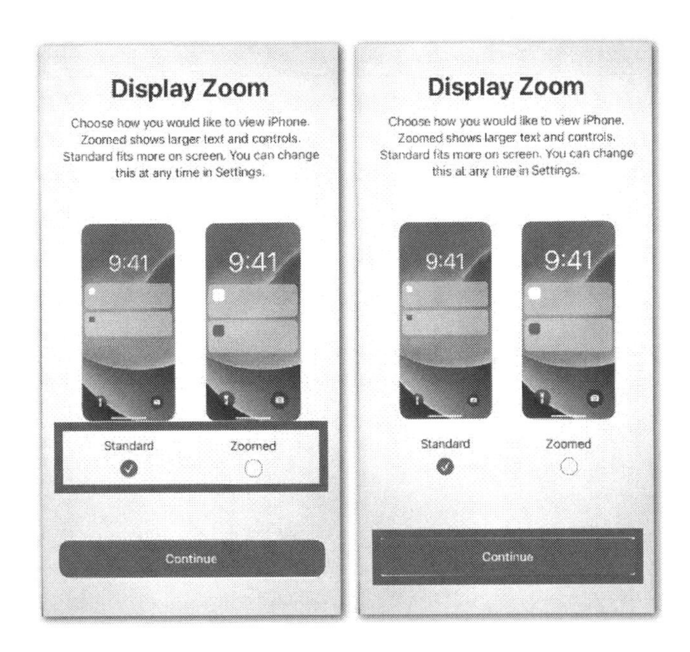

21. Swipe up from the bottom of the screen to get started.

How to Create and Sign in an Apple ID

An Apple ID is an account you use to access all of Apple's services like the App Store, Apple Music, iCloud iMessage, FaceTime, and other Apple services. Apple ID authenticates your identity as an Apple product user through the information and preferences collected from you. It also syncs your account with different Apple products, such as iPhone, iPad, Mac, etc., so they can work together seamlessly.

An Apple ID unlocks a world of possibilities, from storing information in the cloud to syncing data across devices to accessing the App Store and iTunes to buying and downloading content.

Sign into your Existing Apple ID

1. Go to **Settings**.

2. Tap **Sign into your iPhone** under the Search bar.

3. Enter your Apple ID and password.

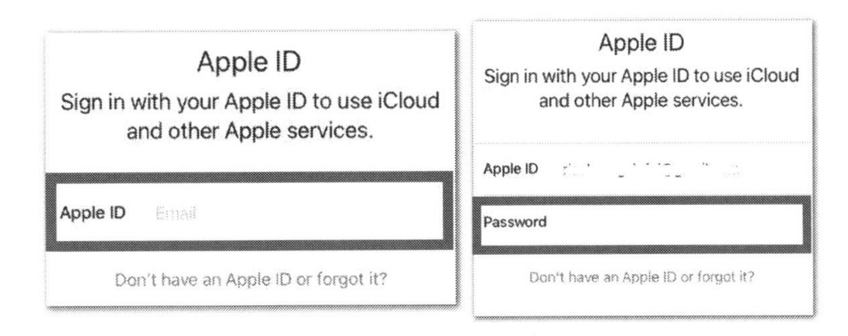

4. If prompted, enter the six-digit verification code sent to your trusted device or phone number and complete the sign-in prompts on the screen.
5. If you have forgotten your Apple ID password, tap **Don't have an Apple ID or forgot it,** then tap **Forgot Apple ID?**

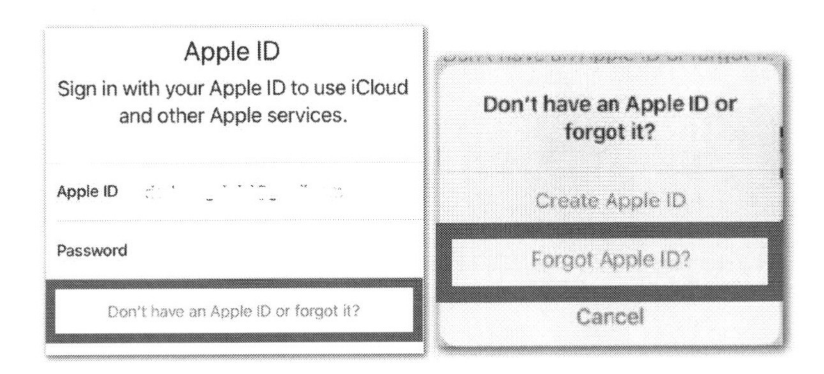

6. Enter your Apple ID, then tap **Next** at the top-right corner of your screen. Confirm your phone number, then tap **Next** at the top-right corner of your screen.

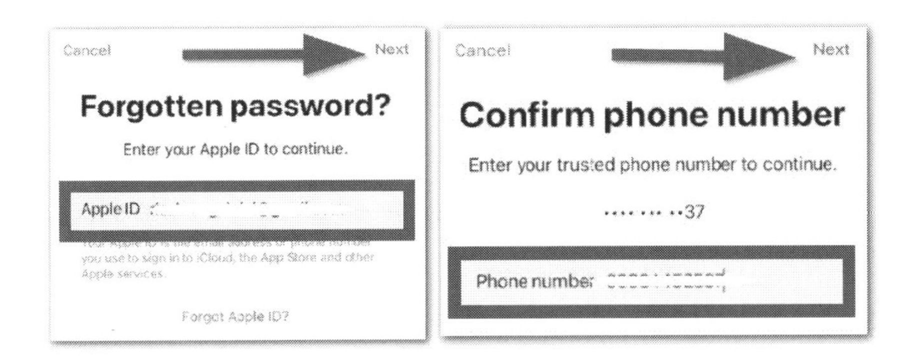

7. Enter the passcode you use to unlock your old iPhone. Enter the New Apple ID password. Follow the instructions to create a password, then tap **Next** at the top-right corner of your screen.

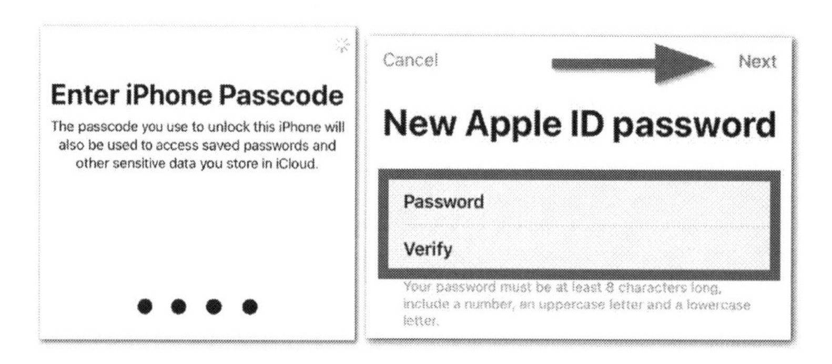

8. Select how you want to keep your data with iCloud. When you have successfully changed you password, tap **Done**.

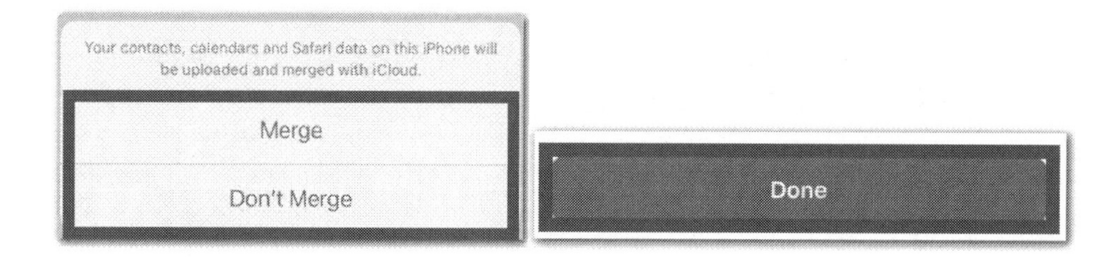

Create an Apple ID

If this is your first Apple product, this guide will help you create an Apple ID in a few simple steps.

1. Go to **Settings**.

2. Tap **Sign in to your iPhone** under the Search bar. Tap **Don't have an Apple ID or forgot it?**

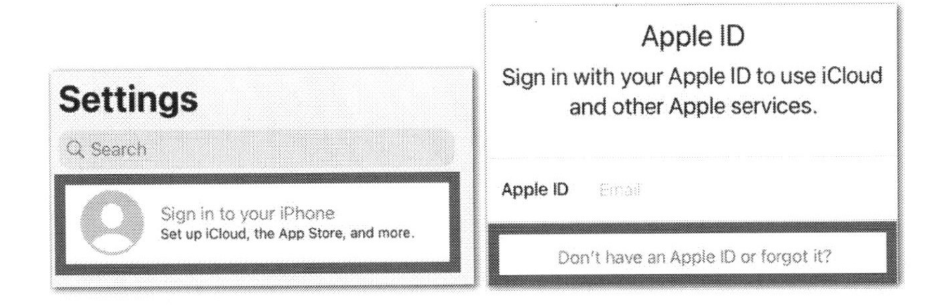

3. Tap **Create Apple ID**.

4. Enter your details, then tap **Continue**.

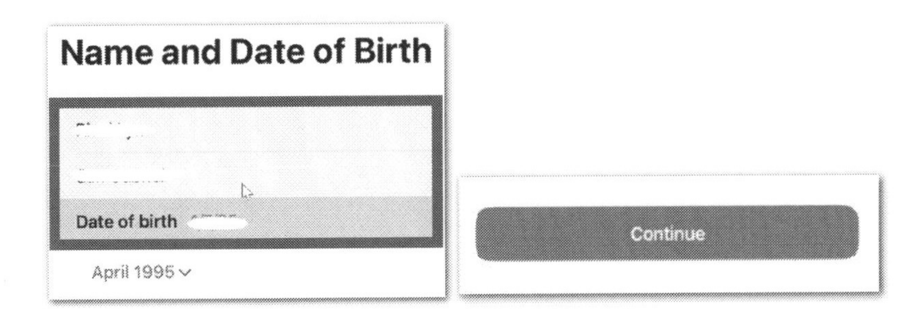

5. Type the **email address** you want to use or tap **Use an existing email address** if you wish to use an existing one.

6. Tap **Continue**.

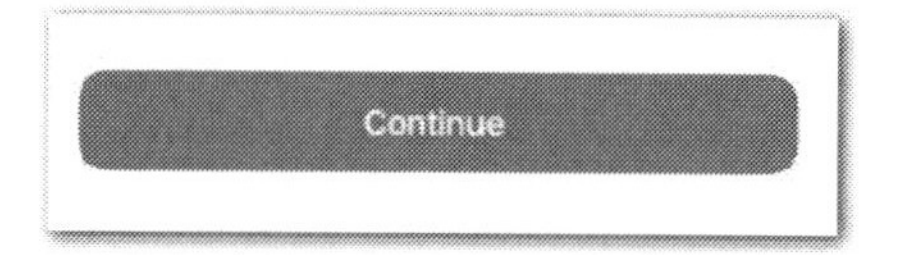

7. Create a **password**, then tap **Continue**.

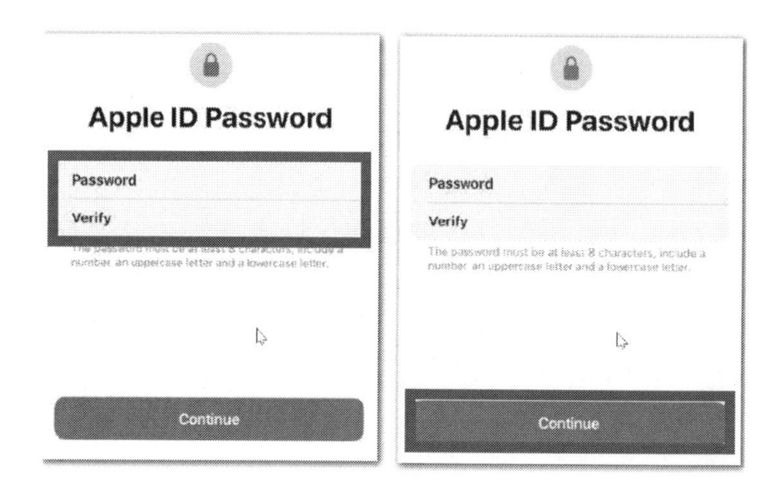

8. Tap Continue if you want to use the phone number on the screen to verify your identity. Tap **Use a different number** if you use another phone number.

9. Tap **Verify email address** to complete your sign-up. After successfully creating and signing into you Apple ID, you can finally use its features.

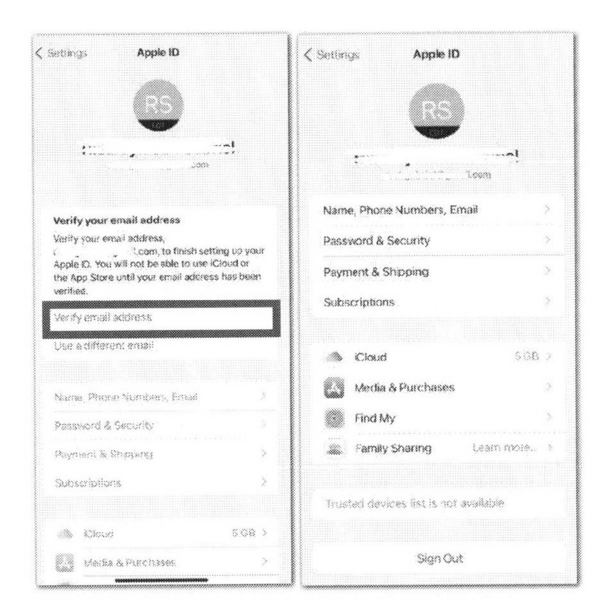

How to Set Up a Passcode

A passcode is a numerical security code that shields your personal information from being accessed by unwanted users. It is a feature available to iPhone to ensure its only sole accessibility for the owner and to prevent cybercriminals from exploiting data stored in it in the event of loss or theft. When you set up a passcode, iPhone enables the auto-lockout feature and your device prompts for your passcode each time you turn it on or wake it up before you can start using it.

On iPhones 13 and 14

1. Go to **Settings**.

2. Scroll down or type **Face ID & Passcode** on the search bar. Tap **Face ID & Passcode**. (You can now go to number 3 below.)

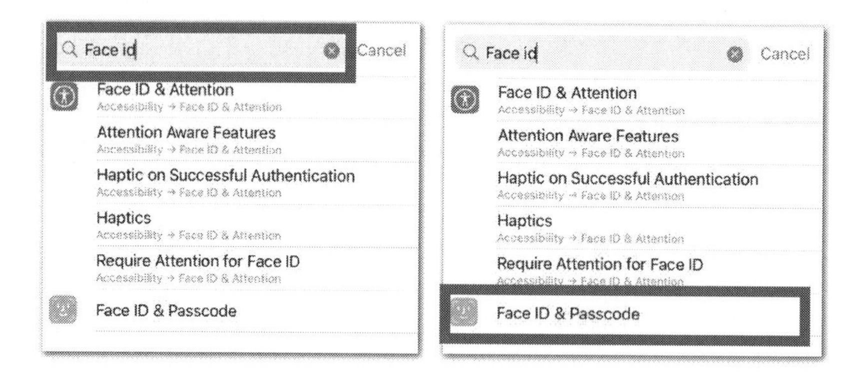

On iPhone SE

1. Go to **Settings**.

2. Scroll down or type **Touch ID & Passcode** on the search bar. Tap **Touch ID & Passcode**.

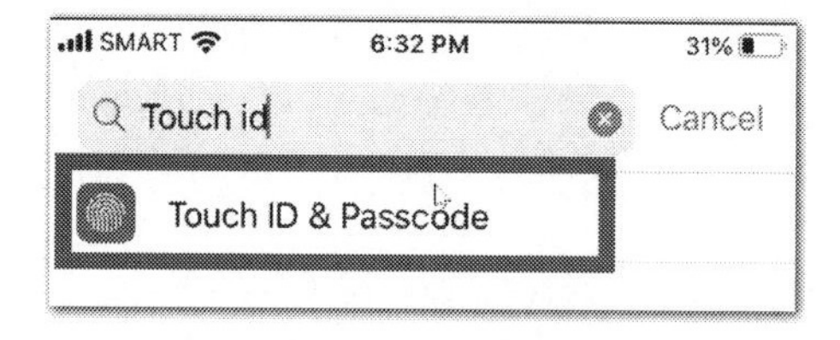

3. Tap **Turn Passcode On**.

4. By default, your iPhone will ask you for a six-digit number. Tap **Passcode Options** at the bottom view the following passcode options:

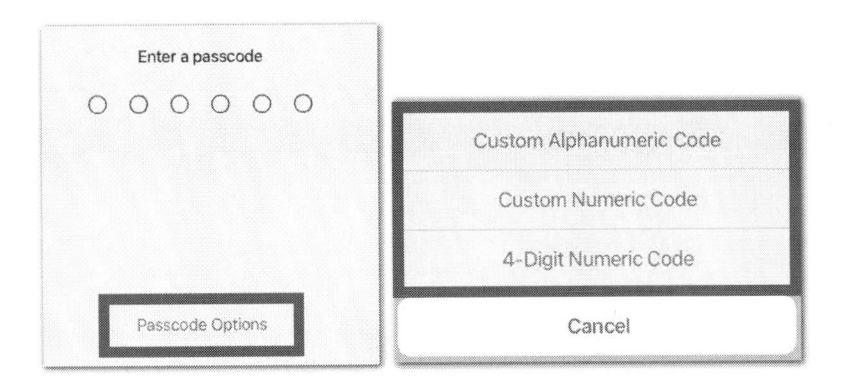

1. Custom Alphanumeric Code – a combination of characters that can be alphabets (upper and lower cases), numbers, symbols, or punctuation marks, e.g., @, $,#.
2. Custom Numeric Code – a combination of multiple numbers
3. 4-Digit Numeric Code – a combination of only four-digit numbers or Change Passcode.

5. Follow the screen prompts, then enter your Apple ID's password.

Tips to Help You Remember your Passwords

Setting passwords on devices is the most secure way to protect personal information and keep it private. Unfortunately, passwords are often forgotten at some point in many people's lives. This scenario poses a massive problem to data security and must be prevented strictly. Refer to the following techniques to better keep your passwords secured:

1. Use the pen and paper method. Write down your passwords on a card, notepad, or journal that you can keep private and inaccessible to anyone. Don't use a piece of paper that can be easily thrown away or lost.

2. Store passwords on other devices, such as computers, tablets, or mobile phones. Save them in notes, spreadsheets, or document-writing apps on these devices. Make sure that these gadgets are protected by passwords that you can safeguard.
3. Use password managers on your mobile or computer. Password managers are applications that store and organize your passwords in one place. Tip: Subscribe to paid password applications for credible security.

How to Set Up Touch ID (iPhone SE) and Face ID (iPhone 13, 14, 15)

Touch ID and Face ID are biometric verification features that Apple introduced to iPhone and its othe products to verify a user's identity, unlock device, authorize payment via Apple pay and make purchase in Apple digital stores. Popular banking and social networking apps also rely on these security features a an extra layer of security in addition to passcodes. Both provide fast access to your phone and authoriz payments for your convenience while protecting your data from threats. This technology uses physica characteristics to verify the identity of anyone who will attempt to access your device. Only the registere person(s) (for Face ID) or fingerprint(s) (for Touch ID) is allowed to unlock your device and create action on it.

Set up Touch ID on iPhone SE

iPhone SE has a Touch ID feature that registers any of your fingers when you place it on the Home butto to confirm your identity. You can add more fingerprints and name them using this feature.

1. Go to **Settings**.

2. Scroll down or type **Touch ID & Passcode** on the search bar. Tap **Touch ID & Passcode**.

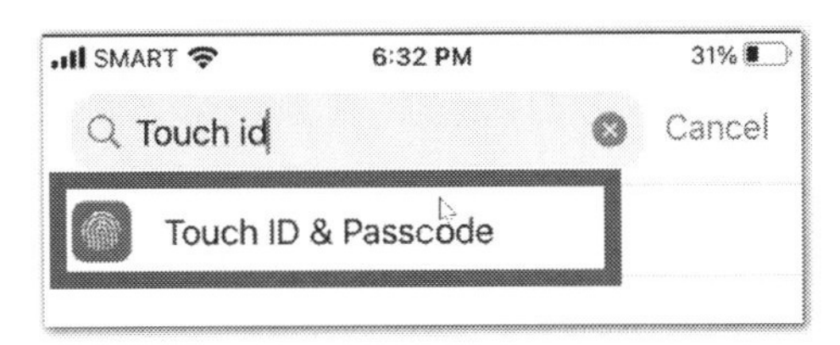

3. Enter your passcode.

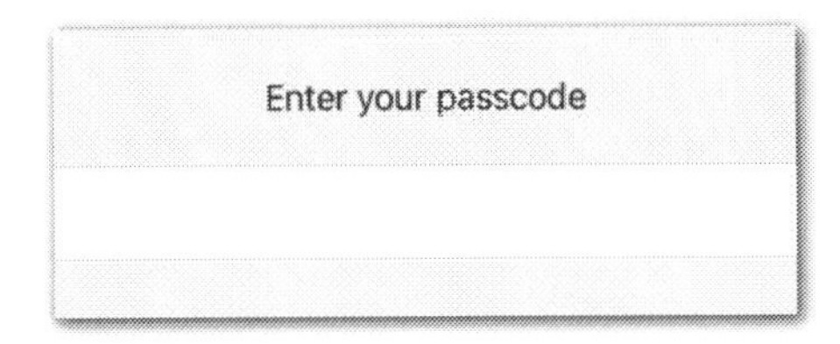

4. Tap **Add a Fingerprint**. Follow the prompts on the screen.

'ou can add several other fingers to Touch ID. Name each fingerprint or delete one to remove access.

'o add more fingerprints:

1. Go to **Settings**.

2. Scroll down or type **Touch ID & Passcode** on the search bar. Tap **Touch ID & Passcode**.

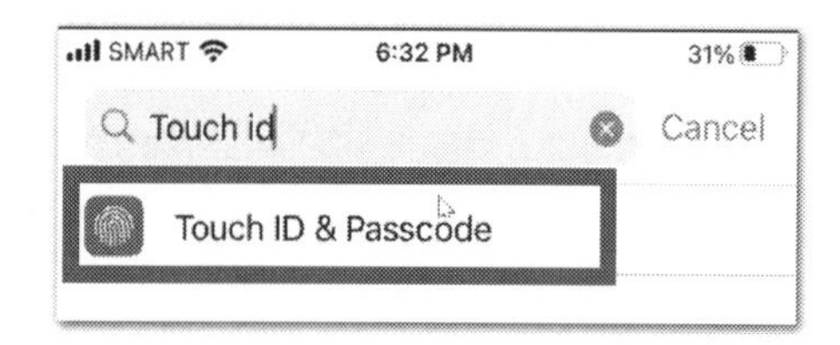

3. Tap **Add a Fingerprint**. Follow the same prompts on the screen.

To name or delete a fingerprint:

1. Go to **Settings**.
2. Scroll down or type **Touch ID & Passcode** on the search bar. Tap **Touch ID & Passcode**.

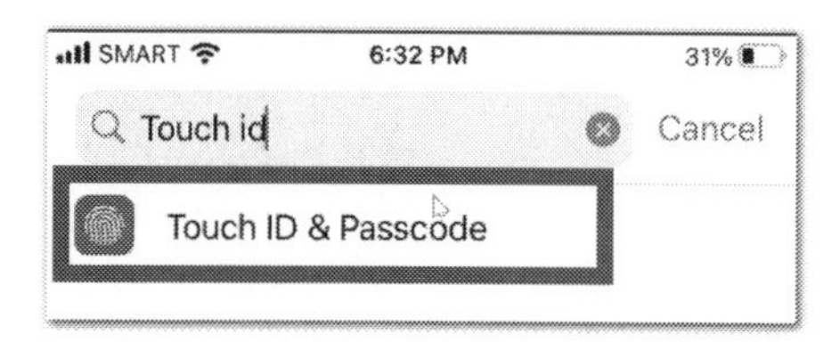

3. Press a finger on the **Home button** to detect its fingerprint among the registered fingerprints.

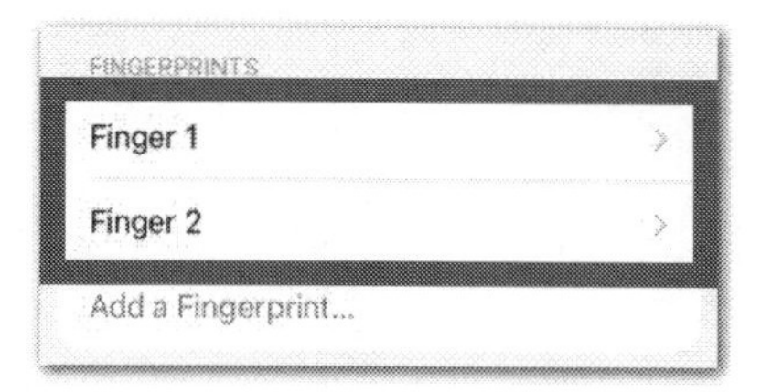

4. Select the fingerprint. Type a name or tap **Delete Fingerprint**.

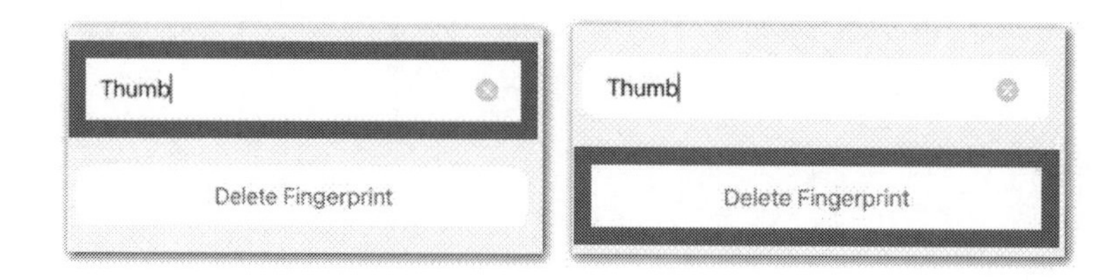

Set up Face ID on iPhone 13 and 14

Face ID is the latest and most advanced security feature for iPhones, using face recognition to verify a person's identity. Similar to Touch ID, you can add another appearance to Face ID.

1. Go to **Settings**.

2. Scroll down or type **Face ID & Passcode** on the search bar. Tap **Face ID & Passcode**.

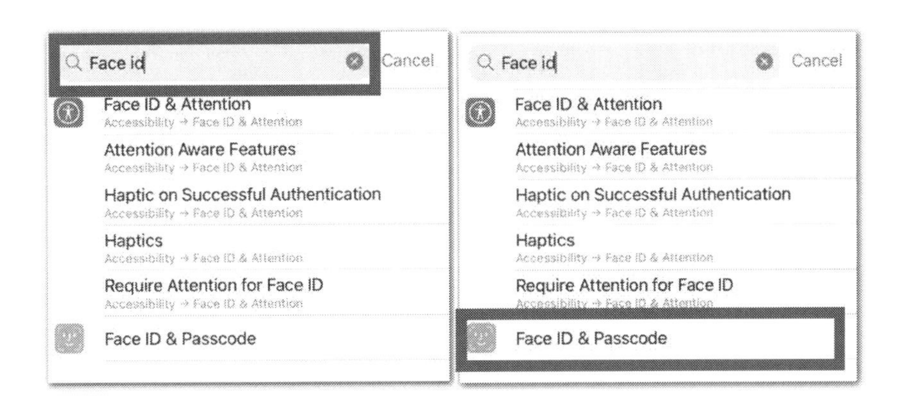

3. Scroll down and tap **Set Up Face ID**. Follow the prompts on the screen.

To add another Face ID:

1. Go to **Settings**.

2. Scroll down or type **Face ID & Passcode** on the search bar. Tap **Face ID & Passcode**.

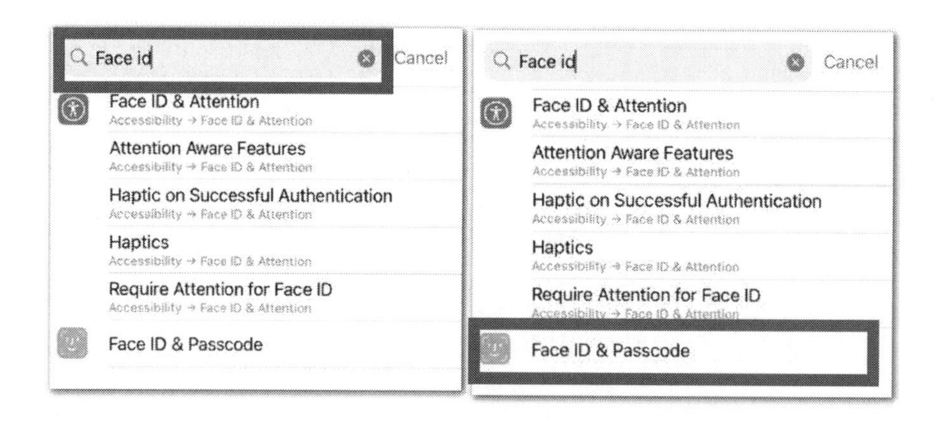

3. Tap **Set Up an Alternate Appearance**. Follow the prompts on the screen.

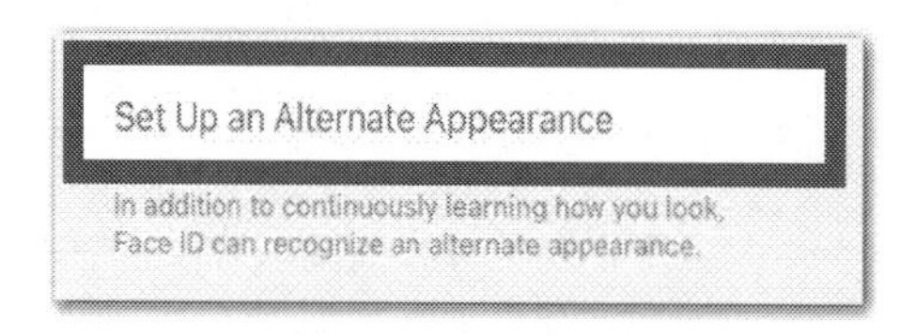

If you are visually impaired or have poor vision, Face ID has an accessibility feature that will not requir you to look at your device with your eyes open. If you enable VoiceOver during the phone setup, this automatically disabled.

1. Go to **Settings**.

2. Scroll down or type Accessibility on the search bar. Tap **Accessibility**.

3. Tap **Face ID & Attention**. Switch off **Require Attention for Face ID**.

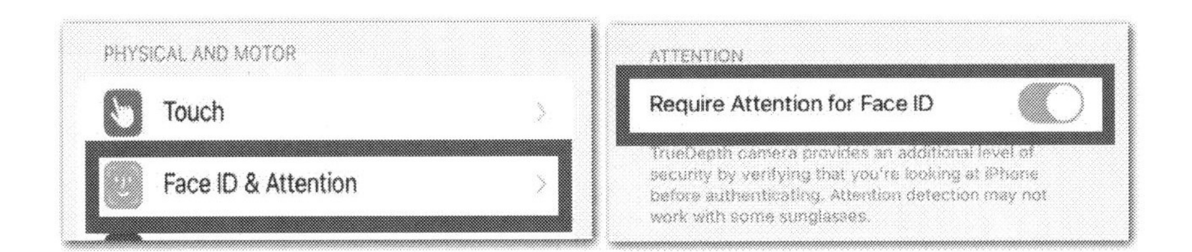

CHAPTER 03

THE BASICS

In this chapter, you will be able to learn the basics of using your iPhone after the complete setup. One of the basic menus in iPhones includes the Control Center or the area where you can find the most used feature in your device. To seamlessly access your phone apps and menus, you will also learn basic navigation. Personalizing your device is also added, such as keyboard, display, text, and language setup, so you can enjoy your iPhone with ease.

Control Center

iPhone makes your phone experience more accessible and enjoyable. Thanks to the Control Center, you can access frequently used features on your iPhone with just one tap. You can change the Wi-Fi settings, take a screenshot, turn on or off Airplane Mode, control volume or display brightness, and more. In other words Control Center is the shortcut to the essential tools on your device. In this chapter, you will understand how to efficiently use Control Center and be more comfortable using most of the iPhone's features. To view the Control Center, swipe down from the top-right corner of your screen (for iPhones 13 and 14) or swipe up from the bottom (for iPhone SE).

Network

The Network tile lets you manage your phone's network activity. It consists of icons, such as airplane mode, cellular data, Wi-Fi, Bluetooth, AirDrop Receiving, and Personal Hotspot. Push and hold the panel for a full view.

1. **Airplane mode** disables all network connections on your iPhone but still enables you to use Bluetooth or Wi-Fi.

1. To enable or disable Airplane mode, tap the **Airplane icon**. The airplane mode is disabled when this option is grayed out. If it turns orange, it signifies you're in Airplane mode.

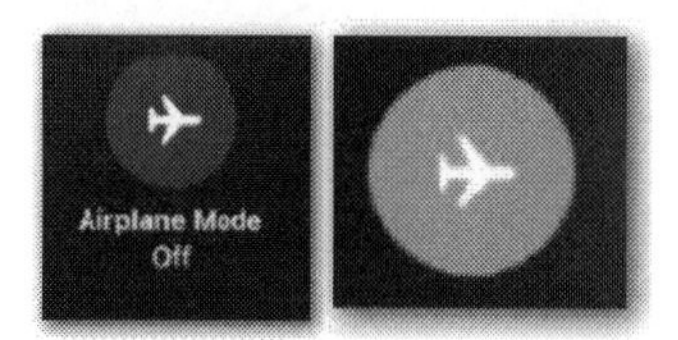

2. You can use either a cellular data network or a Wi-Fi network to connect to the Internet on your smartphone. Cellular **data** operates on the signal coming from the cell towers. It is provided by your carrier and usually has a monthly data cap that you pay for. When you subscribe to a specific data size, you need to track your internet usage to fit in the data allocated to your phone. Otherwise, you will need to pay for any surcharges. The upside of using cellular data is accessibility whenever you are inside your carrier's data coverage. The tower or cellular data icon is used to turn on cellular data.

    1. If you don't want to use your Wi-Fi, tap the **tower icon** to enable cellular data. This icon will be highlighted in green if turned on and grayed out if turned off.

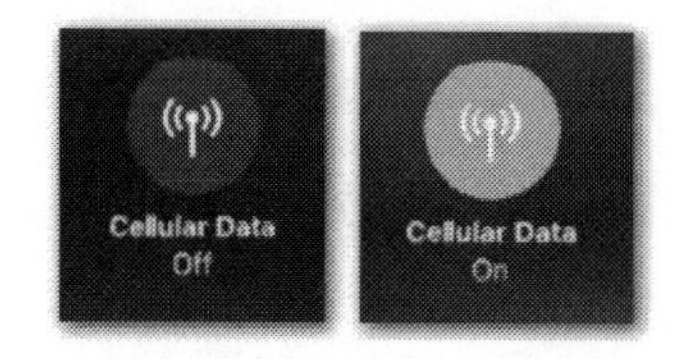

3. **Wi-Fi** transmits the Internet wirelessly to your devices via radio waves and uses a device installed in areas, such as homes, offices, cafes, etc., with a limited range. People love using a Wi-Fi connection because it has no data cap, unlike cellular data. This implies that even if you use Wi-Fi all the time there will never be extra charges on top of your internet bill. So, you can enjoy large data-consuming activities like streaming videos, playing mobile games, and downloading apps.

    1. Tap the **Wi-Fi icon**, which turns blue when enabled.

2. Push and hold the **Wi-Fi icon** to change to a different Wi-Fi network.
3. The list of available Wi-Fi networks will then appear, tap on the **network** to connect.

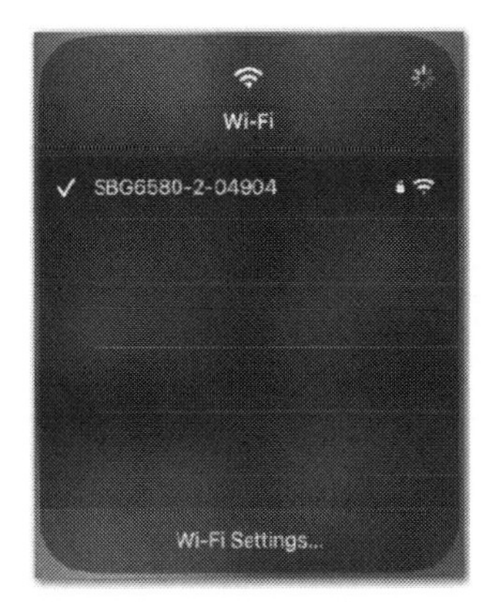

4. **Bluetooth** is a phone feature that wirelessly connects and transfers data across short distances between Bluetooth-enabled mobile devices.

   1. Tap the **Bluetooth icon** to enable Bluetooth.

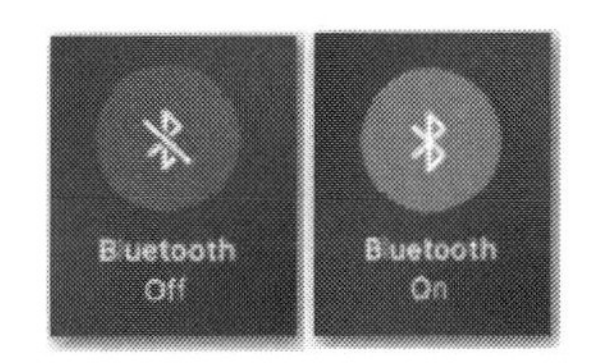

   2. Push and hold the **Bluetooth icon** to see a nearby Bluetooth device, then tap on the **Bluetooth name** to connect.

5. Use **AirDrop** for short-distance data sharing and more with other Apple devices faster than Bluetooth. AirDrop only works with a similar device and cannot be paired with other smartphones (e.g., iPhone to iPhone only).

1. Push and hold the network panel, then tap the AirDrop icon.
2. Select if you want to send or receive from **Contacts only** or **Everyone**. After selecting, the Bluetooth and AirDrop icons will automatically enable and turn blue.

6. **Personal Hotspot** is a technology that enables other devices, such as computers and smartphones, to connect to your cellular data, so they can connect to the internet. Think of Personal Hotspot like your own Wi-fi connection that requires those who wish to connect with you using a password.

    1. Tap the **tower icon** to enable cellular data. Tap the **Personal Hotspot icon** and your network becomes visible to others to connect.

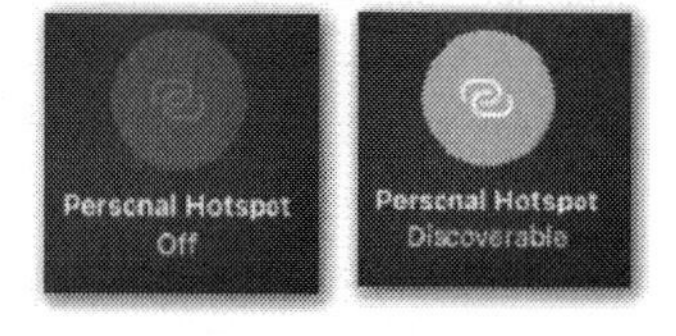

Now Playing

This tile lets you control music or video on your device. With Now Playing, you can play, pause, turn the volume up or down, skip between songs, etc. Push and hold the tile for a full view.

Orientation Icon

With the orientation icon, you can change your phone between portrait and landscape mode. Tap the **orientation icon** to lock your phone to portrait. Tap **unlock** if you want to allow rotation to landscape mode.

Two Boxes Icon

This icon refers to screen mirroring, which copies your phone's screen and shows it to another Apple-supported device like Apple TV. Tap the **two boxes icon** to see the available devices, then tap on the device you want to screen mirror.

Moon Icon

The moon icon enables the previously used "Focus mode." Focus mode helps you concentrate and limit distractions, so you can focus on what you're doing. Push and hold the **moon icon**, then select the available Focus modes.

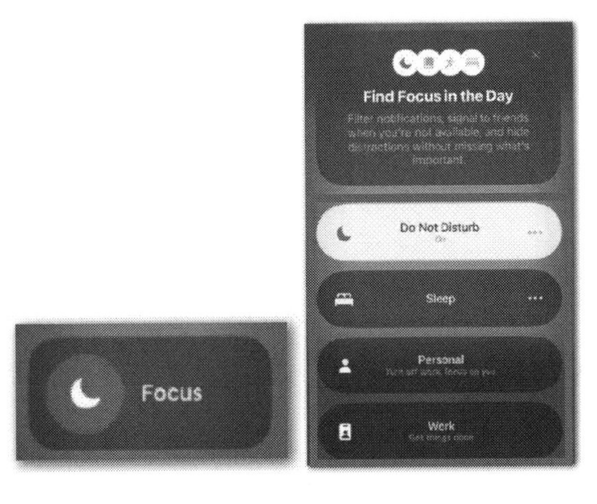

Brightness Slider

If you want to adjust the brightness of your iPhone's display, scroll down on the brightness slider. Push and hold the slider to view other lighting modes, then tap the available icons to change modes.

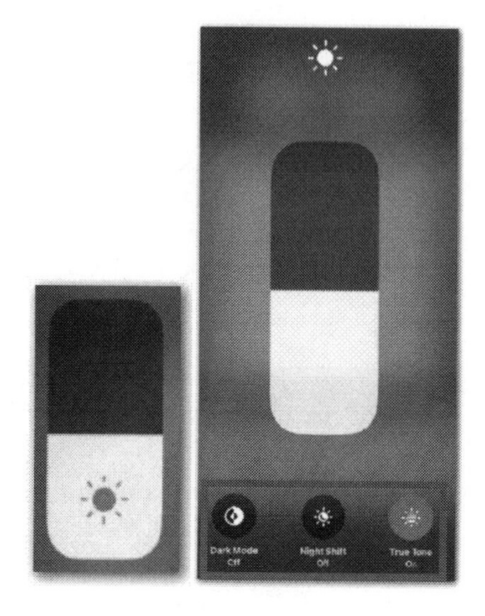

Volume Slider

The volume slider changes your phone's volume. However, it does not affect the ringtone or notification volumes. You can change the ringtone or notification volume using the volume buttons.

Customize Your Control Center

Edit how you want to see your iPhone's control center by adding shortcuts for Flashlight, Screen Recording, Camera, Dark Mode, and many more.

1. Go to **Settings**.

2. Scroll down or type **Control Center** on the search bar. Tap **Control Center**.

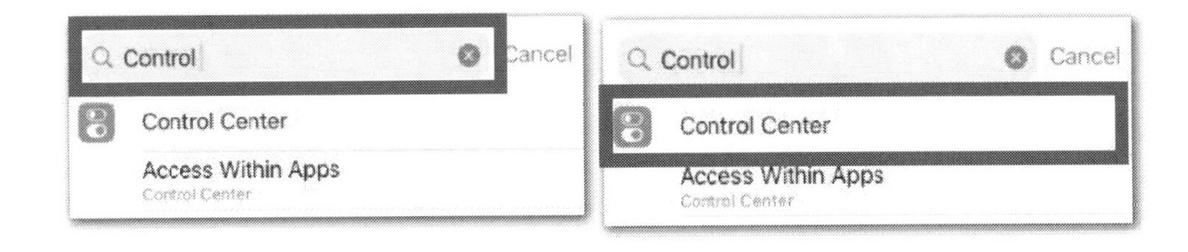

1. Under More Controls, tap the **add button** next to the app you want to add. To remove an app from the Control Center, tap the **delete button** next to the app you want to remove.

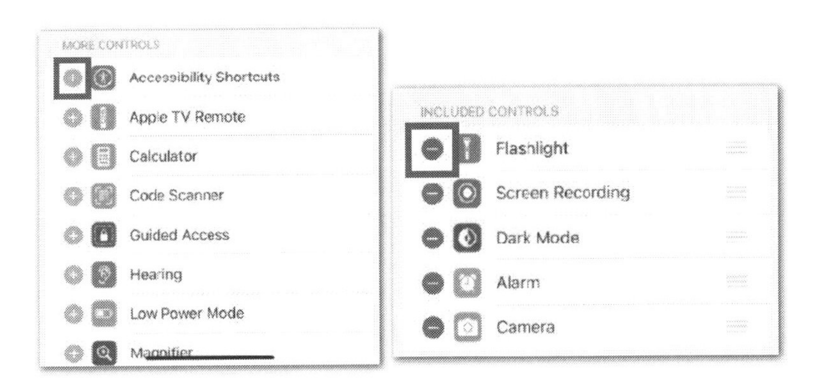

How to Set Up Keyboards

etting up keyboards on your iPhone makes typing convenient with its powerful features, including spell-hecking, smart punctuation, and auto-correction. If you're someone who uses different languages, you an add keyboards available in multiple languages then navigate keyboard languages while typing. You can lso choose different keyboard layouts and enlarge your keyboard size for a bigger typing space.

Set up keyboard preferences

1. Go to **Settings**.

2. Scroll down or type **General** on the search bar. Tap **General**.

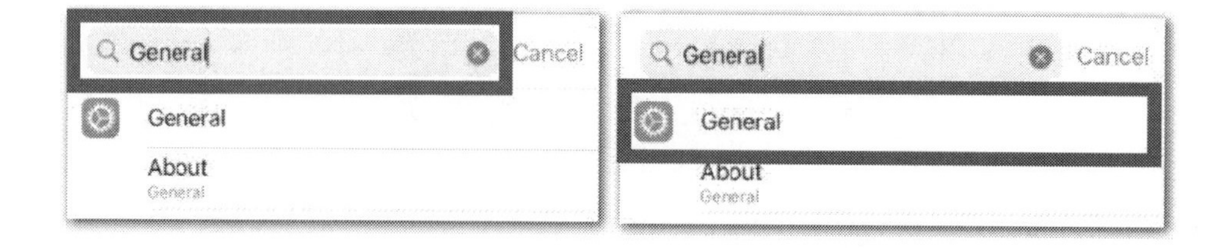

3. Scroll down and tap **Keyboard**. Under All Keyboards, tap any feature you wish to enable.

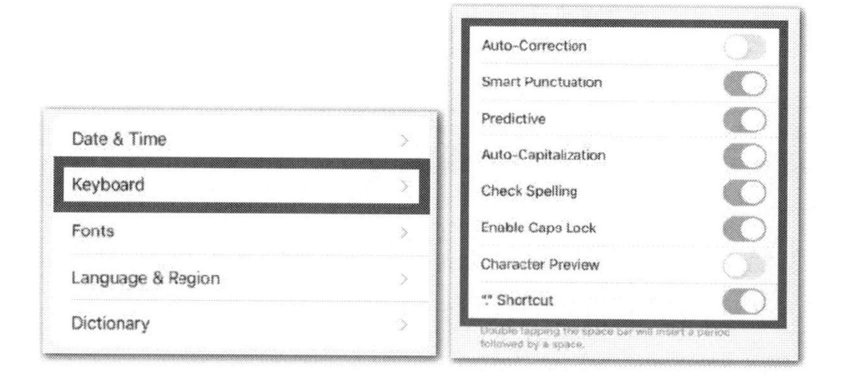

Add a keyboard

1. On the Keyboards screen, tap **Add New Keyboard**. Select a keyboard from the list.

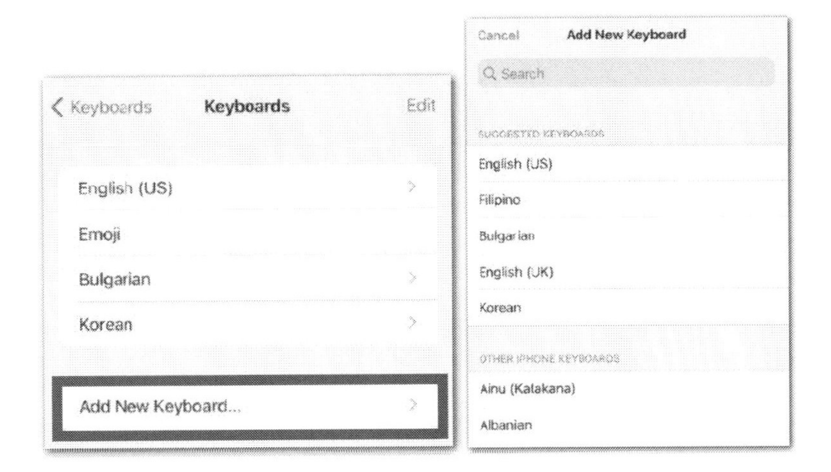

Rearrange your keyboard list

1. On the Keyboards screen, tap **Edit**, then move the **three-bar icon** of the keyboard to a new order in the list. Tap **Done** at the top-right corner of your screen.

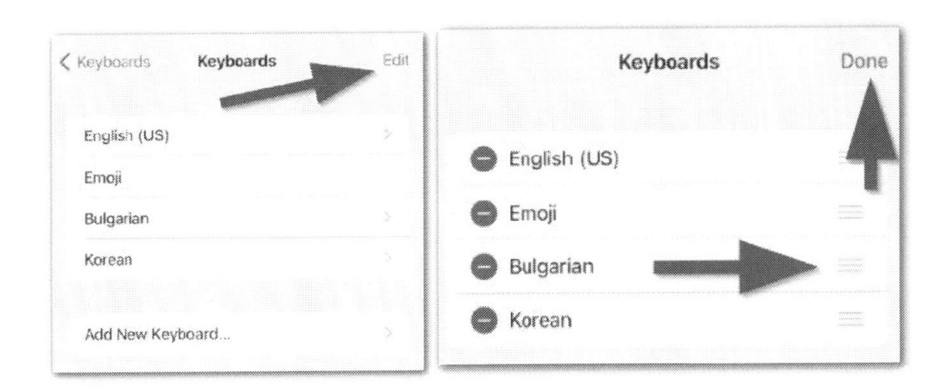

Remove a keyboard

1. On the Keyboards screen, tap **Edit**, then tap the **delete icon** next to the keyboard language. Tap **Done** at the top-right corner of your screen.

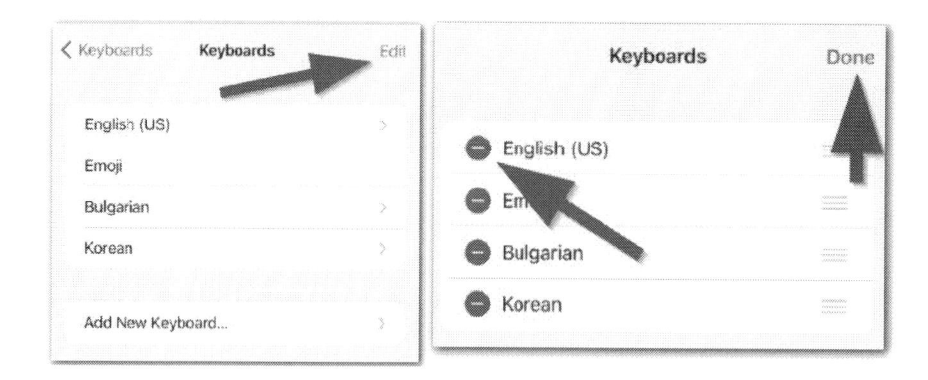

Change a keyboard layout

1. Changing a keyboard layout shows a different order of the keys on your keyboard. On the Keyboards screen, tap a **keyboard**, then select your preferred layout from the list.

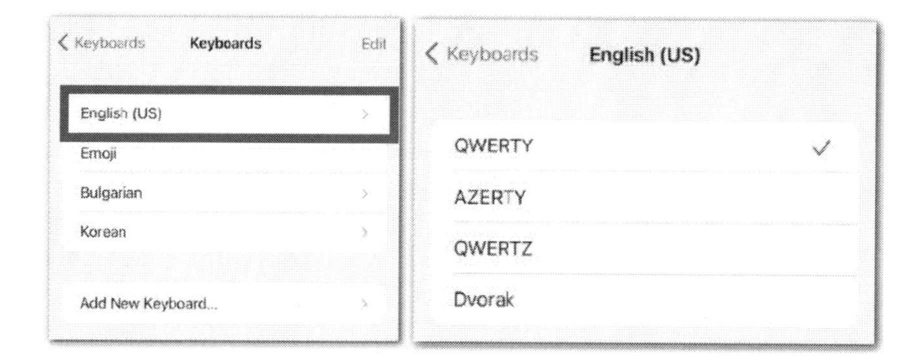

Switch keyboard languages

1. On the typing field, push and hold the **globe icon** at the bottom. Select the keyboard you wish to switch to.

2. You can also tap the globe icon once until another keyboard appears.

Enlarge keyboard screen

1. Scroll down from the top-right corner of your screen to go to Control Center.

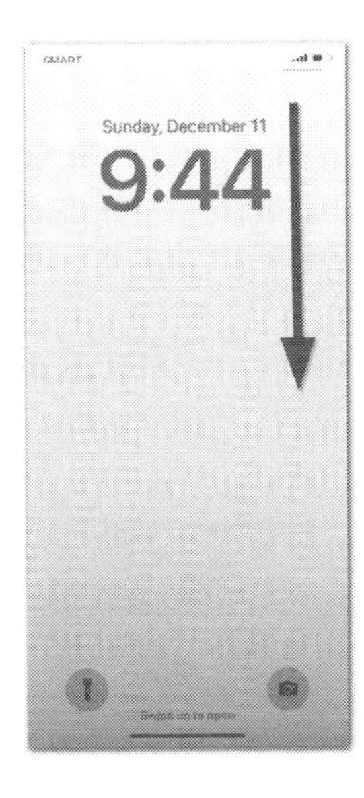

2. Tap the lock icon to unlock.

3. On the typing field, rotate your device to landscape orientation.

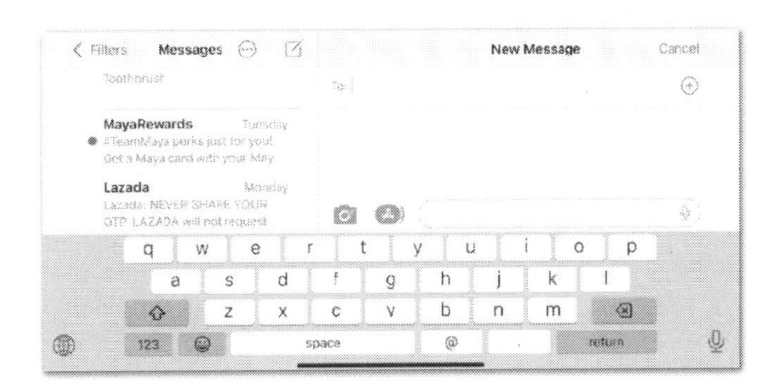

How to Add a Language

After adding a keyboard for another language to your device, you can also opt to add a language to your phone as its main language. Apple supports adding multiple languages and lets you change the main language of your phone easily whenever you need to. Changing the main language of your phone changes the text in most menus and apps. It also automatically changes the time zone and location, which is perfect when you move or travel.

Add a Language

1. Go to **Settings**.

2. Scroll down or type **General** on the search bar. Tap **General**.

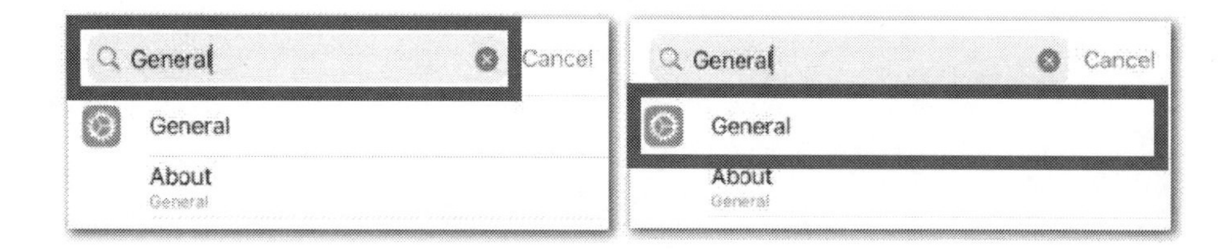

3. Tap **Language & Region**.

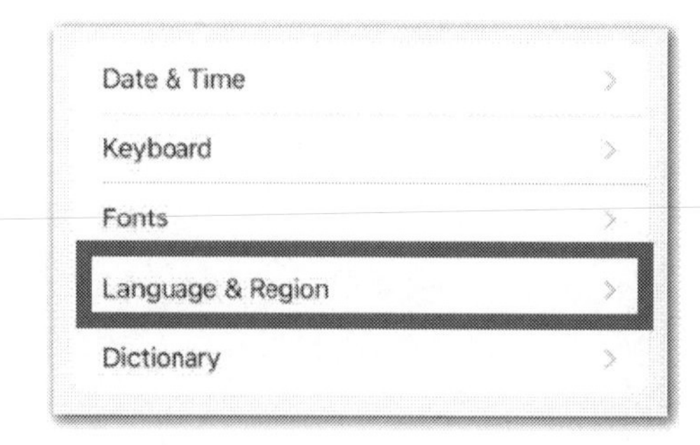

4. Tap **Add Language**. Select a language from the list.

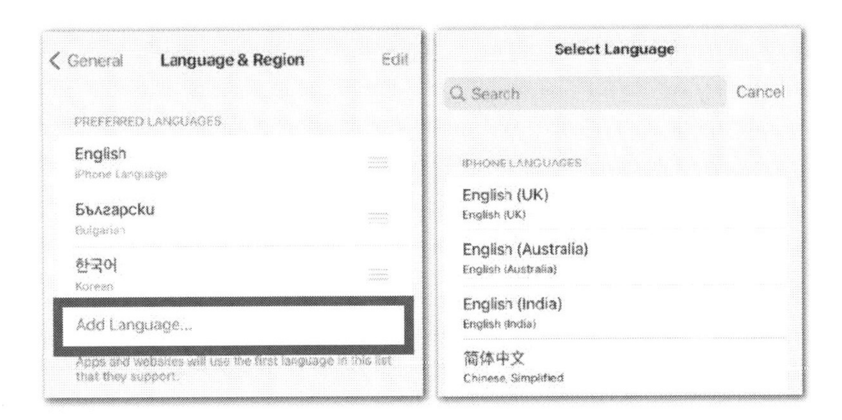

5. Select the language to set it as your phone's primary language and a restart will prompt. Otherwise, select **Cancel**.

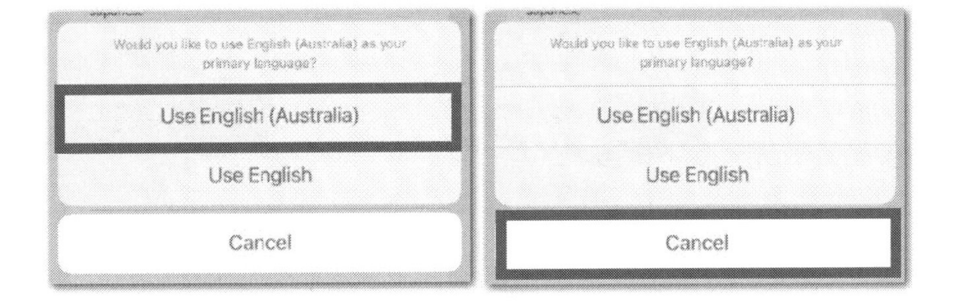

Change your main language

1. From the list of preferred languages, move the three-bar icon of the language to the top. Tap **Continue**. Your phone will restart.

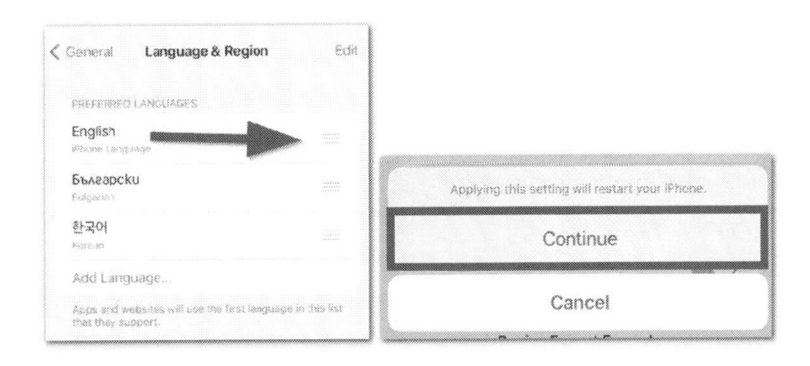

Appearance on iPhone

The iPhone has a variety of screen options, allowing you to use your device in a way that best suits you vision while also customizing it with a design of your choice. Adjusting the screen display's brightness switching between light or dark mode, and changing text size are ways to set your phone's appearance Setting these preferences gives you a comfortable viewing experience and lessens harm to your vision. Plus you can do more than just adding a simple wallpaper and lock screen with iPhone's custom backgrounds from a gallery with plenty of choices and effects.

Adjust screen brightness

Set the iPhone's display's hue and brightness to your liking by adjusting the brightness and contras of your screen display. You can select from Dark Mode, True Tone, and Night Shift which allow you to change the display's brightness and color manually or automatically.

A. Manually adjust brightness or darkness

1. Go to the **Control center**, tap the **Brightness icon,** and slide to adjust brightness.

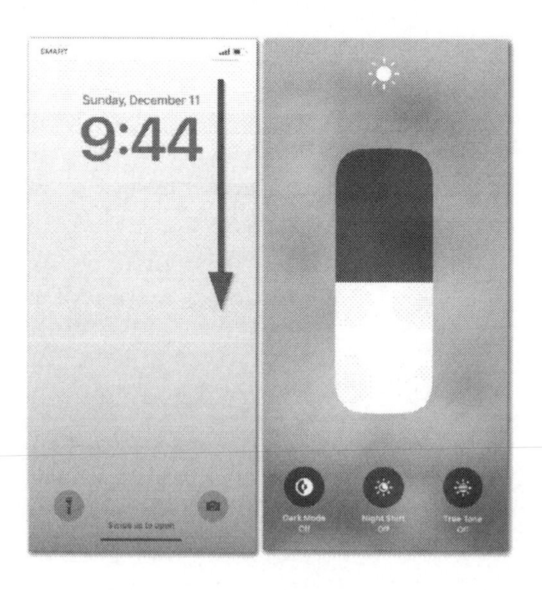

2. Go to **Settings**.

3. Scroll down or type **Display & Brightness** on the search bar. Tap **Display & Brightness**.

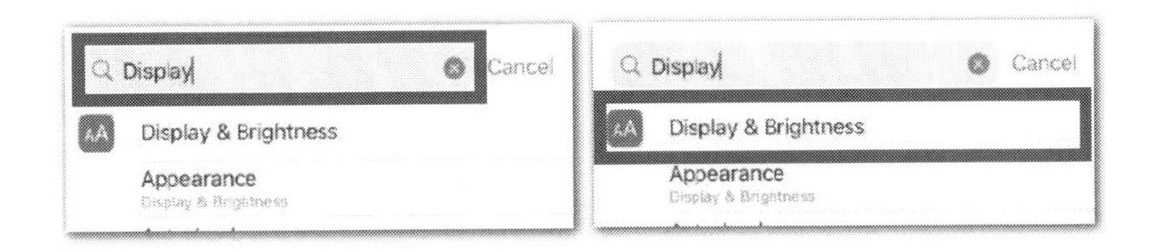

B. Automatically adjust brightness or darkness

1. Go to **Settings.**

2. Scroll down or type **Accessibility** on the search bar. Tap **Accessibility.**

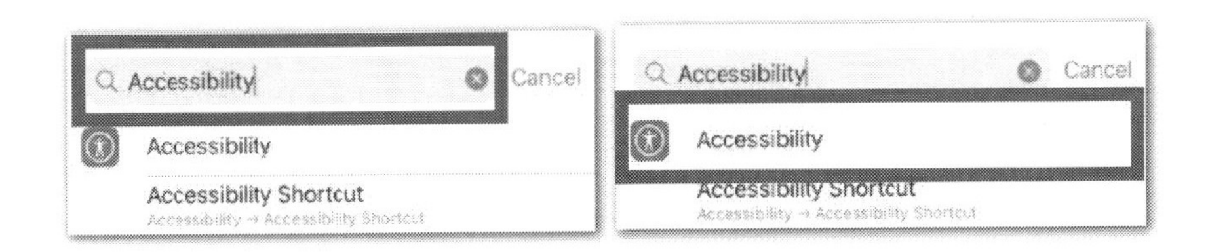

3. Go to **Display & Text Size**, then switch on **Auto-Brightness**.

Turn Dark Mode on or off

With Dark Mode on, your entire iPhone experience will take on a subdued, nighttime aesthetic. Use your iPhone in bed without disturbing your companion by switching to Dark Mode.

1. Go to **Control Center**, then push and hold the **Brightness icon**.

2. Select the **Dark Mode** switch to turn it on or off.

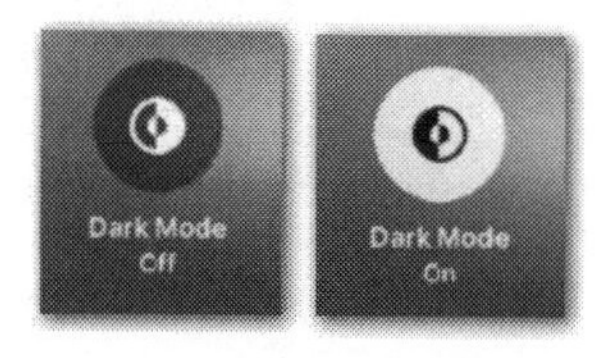

3. Alternatively, you can go to **Settings**.

4. Then, scroll down or type **Display & Brightness** on the search bar. Tap **Display & Brightness**.

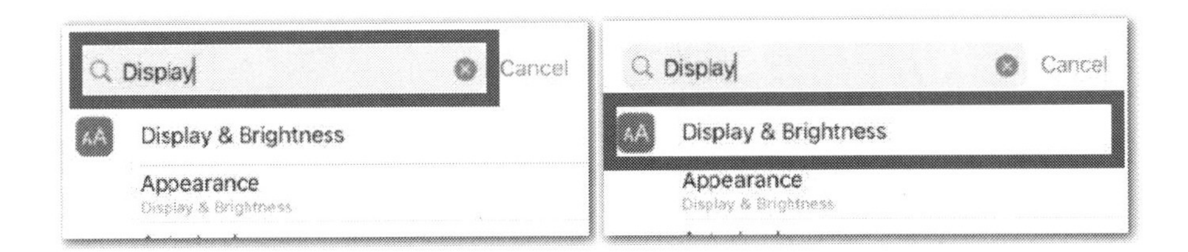

5. Select **Dark** to turn on Dark Mode or **Light** to turn it off.

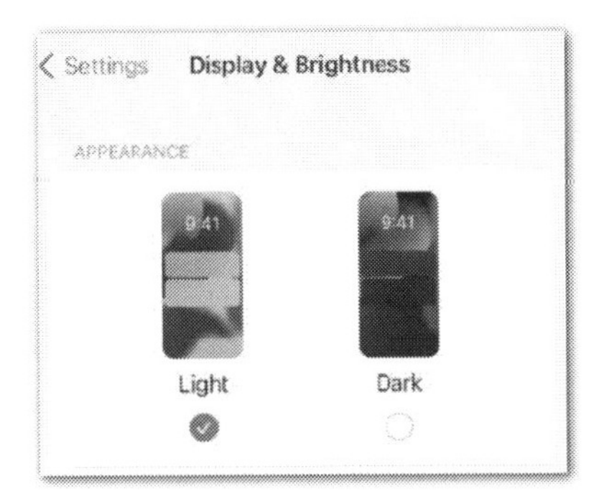

Turn True Tone and Night Shift on or off

True Tone adjusts brightness automatically to better adapt to the ambient lighting, while the Night Shift node shifts the color of your display to the warmer end of the color spectrum after dark. To turn True Tone ınd Night Shift on or off:

1. Go to **Control Center**, then push and hold the **Brightness icon**.

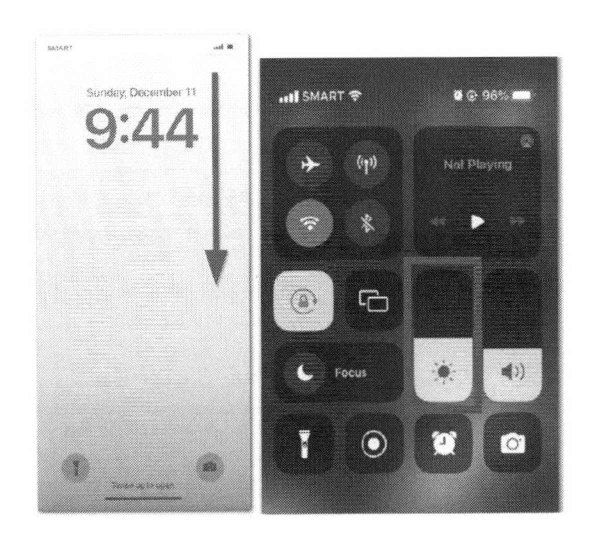

2. Select **True Tone** and **Night Shift** to turn on or off.

3. Alternatively, you can go to **Settings**.

4. Then, scroll down or type **Display & Brightness** on the search bar. Tap **Display & Brightness**.

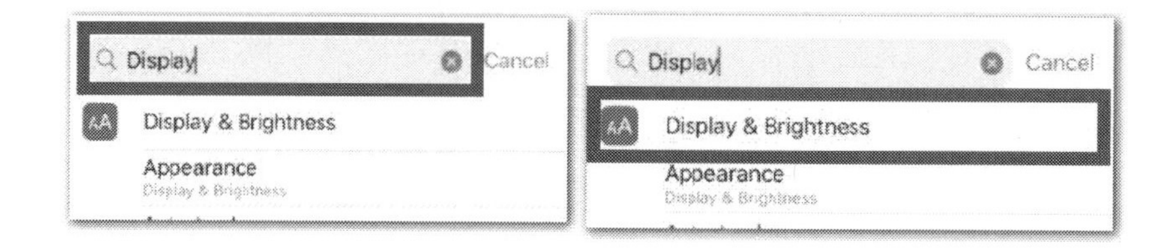

5. Turn **True Tone** and **Night Shift on or off**.

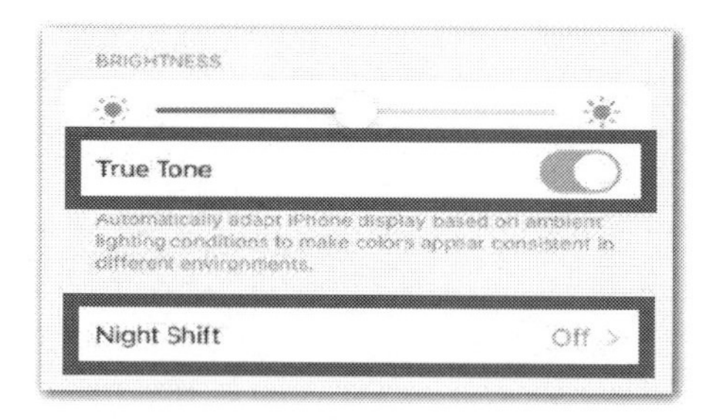

Make a new iPhone wallpaper

Both the Lock Screen and the Home Screen wallpaper can be customized on an iPhone. You can sele a new background from the Settings menu or the Lock Screen's wallpaper collection. You may modify th iPhone's Lock Screen to your liking.

1. Go to **Settings**.

2. Scroll down or type **Wallpaper** on the search bar. Tap **Wallpaper**.

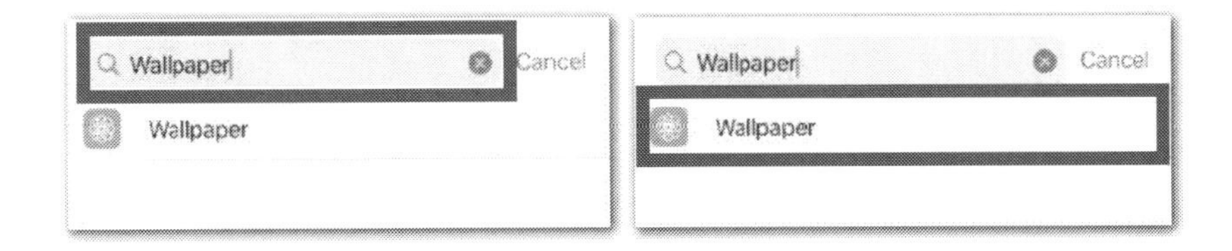

3. Tap **Add New Wallpaper**. When the background gallery loads, select from Photos, People, Photo Shuffle, Emoji, or Weather icons at the top of the wallpaper galleryy. You can design your wallpaper with a photo, an emoji pattern, a picture of your local weather, and more.

4. You can also select a wallpaper option in one of the categories like Featured, Suggested Photos, Photo Shuffle, and so on.
5. Tap **Add** at the top-right corner of your screen. Tap **Set as Wallpaper Pair** if you prefer to use the wallpaper on both the Lock Screen and Home Screen.

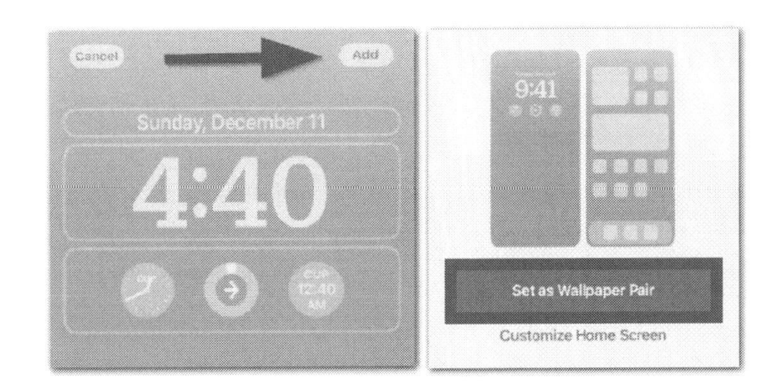

6. If you wish to add elements and modify the Home Screen, tap **Customize Home Screen**. Change the wallpaper's color by selecting a color, a custom photo by tapping the **picture icon**, or **Blur** to make the wallpaper apps stand out

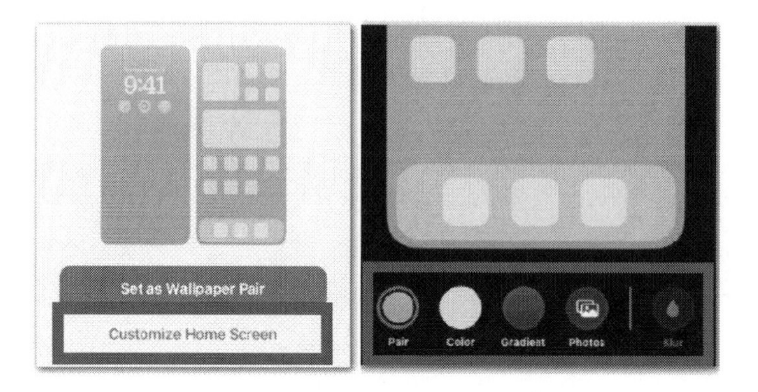

7. If you're happy with your lock screen, tap **Done** at the top-right corner of your screen.

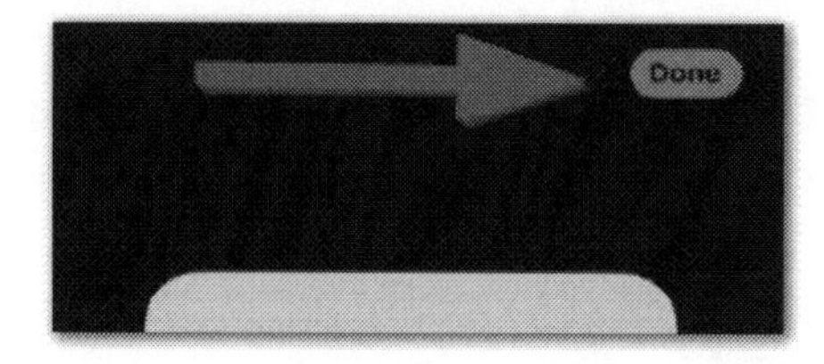

Personalize your iPhone Lock Screen

Lock Screen customization options include changing the wallpaper, altering the color scheme and fonts adding a photo overlay with the time, and many others. The Lock Screen can also host widgets displayin information from your installed apps, such as the current weather, upcoming appointments, and new headlines.

Make as many different Lock Screens as you like, and switch between them whenever you choose. Eac Lock Screen can be associated with a different Focus, so you can easily change your Focus just by selectin a new Lock Screen.

Tip: Setting up Face ID (on a model with Face ID) or Touch ID (on a model with Touch ID) beforehan will make creating a personalized Lock Screen much simpler. Check out the instructions for establishir biometric authentication with either Face ID or Touch ID.

Create a custom lock screen directly from the Lock Screen

1. To access the customization options, push and hold the **Lock Screen** until the button appears at the bottom.

2. If the button to **Customize** the Lock Screen doesn't appear, push, and hold it again, and this time entering your passcode.
3. Tap the **add button** at the bottom-right of the screen, then select a wallpaper. Swiping left or right allows you to explore complementary color filters, patterns, and typography for certain wallpaper options.

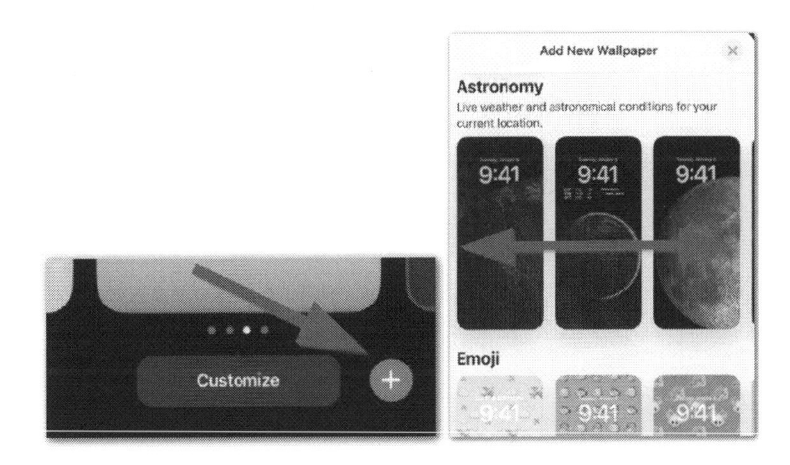

4. Tap **Add**. If you prefer to use the wallpaper on both the Lock Screen and Home Screen, tap **Set as Wallpaper Pair**.

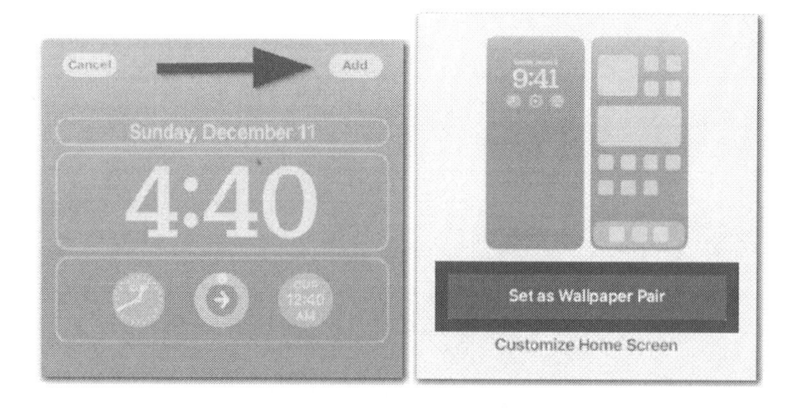

5. To add elements and modify the Home Screen, tap **Customize Home Screen**. Select a color to change the wallpaper color, a custom photo by tapping the **picture icon**, or tap **Blur** to blur and make the wallpaper apps stand out.

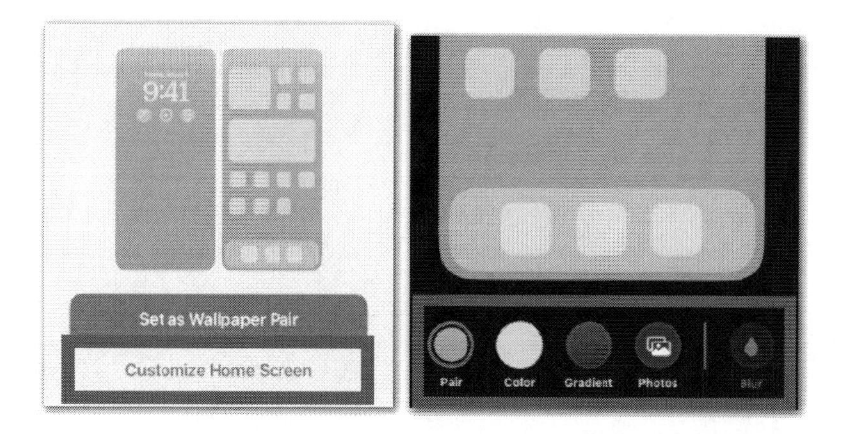

6. If you're happy with your lock screen, tap **Done** at the top-right corner of your screen.

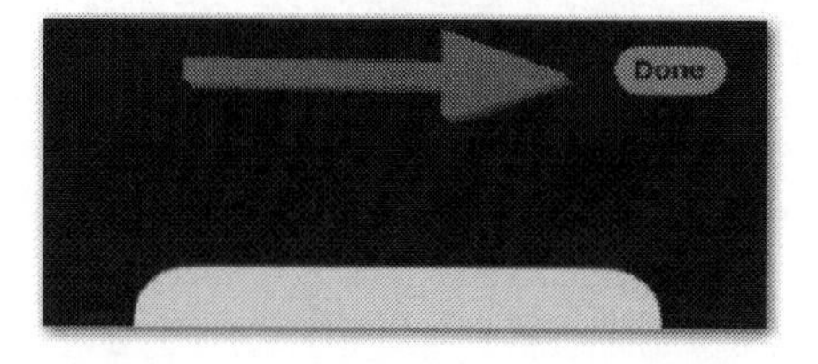

Tip: You may also replace the default images on the Home Screen and Lock Screen with images already stored in your gallery. To view a photo from your device's library:

1. Go to **Photos.**

2. Go to the **Library** tab, then select a photo.

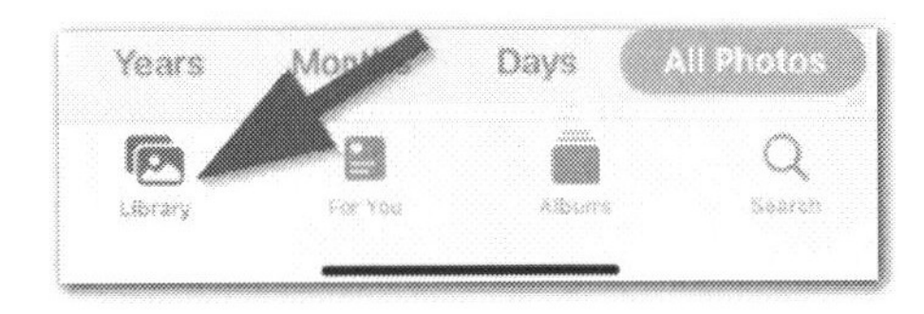

3. Tap the **share button**. Swipe down and select **Use as Wallpaper**.

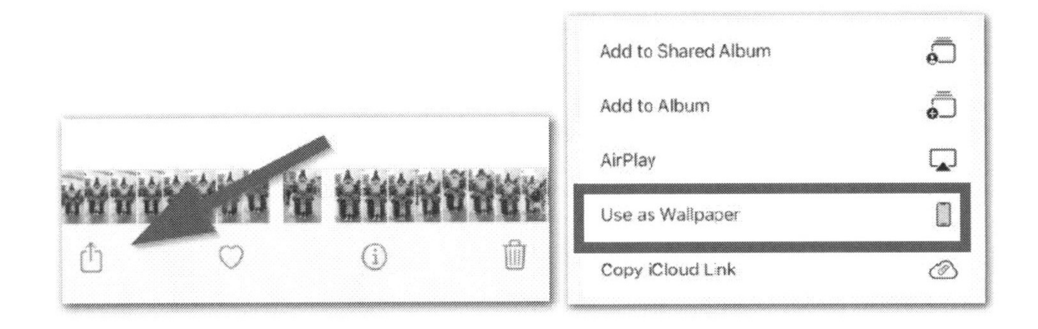

4. You may customize your wallpaper and if happy with your selection, tap **Done** at the top-right corner of your screen.

How to Adjust the Font Size

ou can adjust the screen's brightness, contrast, and other features to help those who are colorblind or ave other visual impairments read the screen. Use display accommodations.

1. Go to **Settings**.

2. Scroll down or type **Accessibility** on the search bar. Tap **Accessibility**.

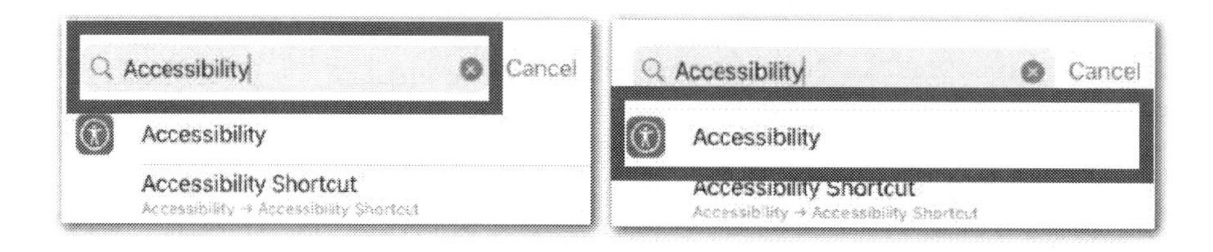

3. Tap **Display & Text Size**.

4. Adjust any of the following:

A. **Bold Text:** Highlight the words and make them stand out.

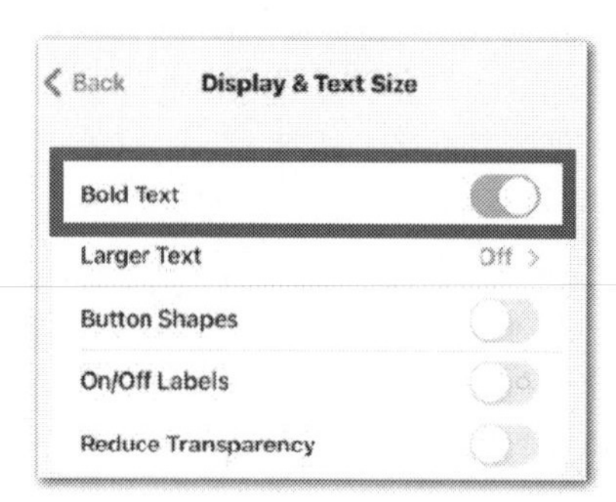

B. **Larger Text**: Makes the text easier to read, enables the Larger Accessibility Sizes setting, and then uses the Font Size setting to make any adjustments.

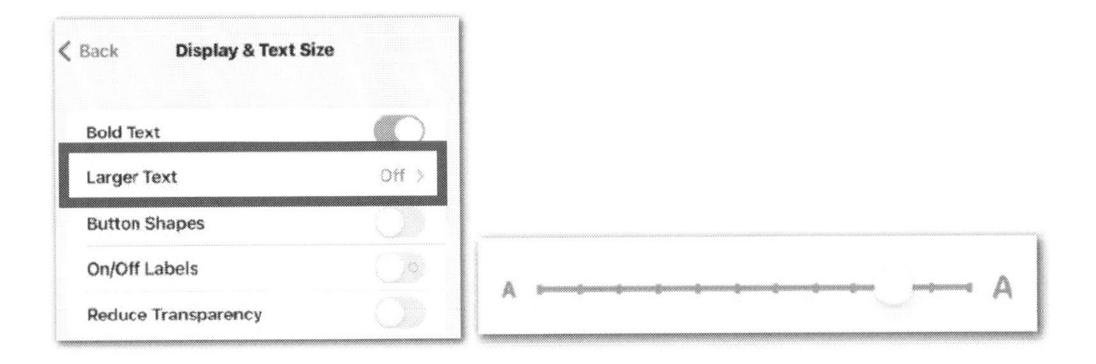

Apps like Settings, Calendar, Contacts, Mail, Messages, and Notes that enable Dynamic Type will adapt to your preferred font size. Adjust the text size when you're using an app.

Basic Navigation

Navigating an iPhone is easy but might be tricky if you are unfamiliar with it. You might wonder how to do a function that does not seem visible on the screen. Don't fret. After learning this topic and constantly navigating your phone, you can become a pro. Here are some simple but helpful navigation tips you should know if you're using an iPhone for the first time.

How to wake up iPhone

Push the **side button** once to turn on the screen. On iPhone SE, you can also push the **Home button/ Touch ID** sensor to wake up the screen.

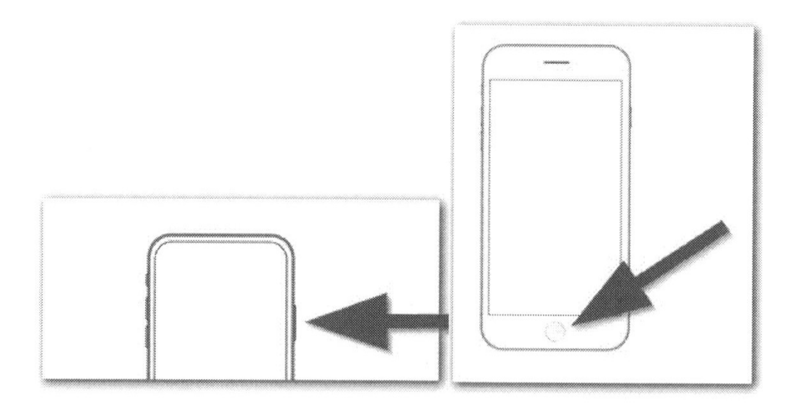

1. On the lock screen, scroll down from the bottom. Enter your passcode if prompted.

2. To turn off the iPhone screen, push the **side button** once.

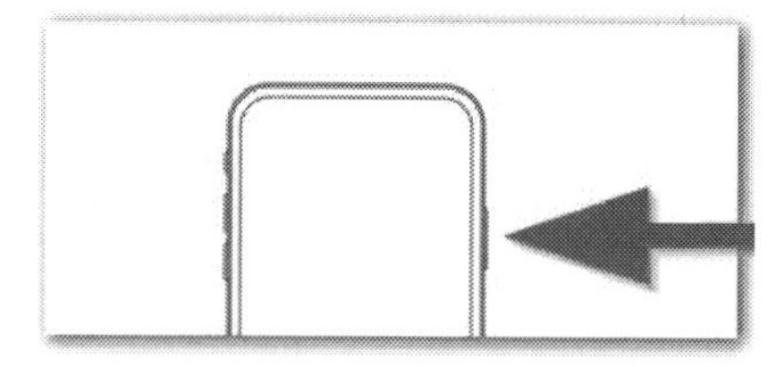

How to go back

1. To go back to the last page on iPhone, tap the **back button** at the top-left corner of the screen. Th button changes depending on the app or the page you're on.

Alternatively, you can swipe from the left of the screen to the right. This also depends on the app or the page you're on.

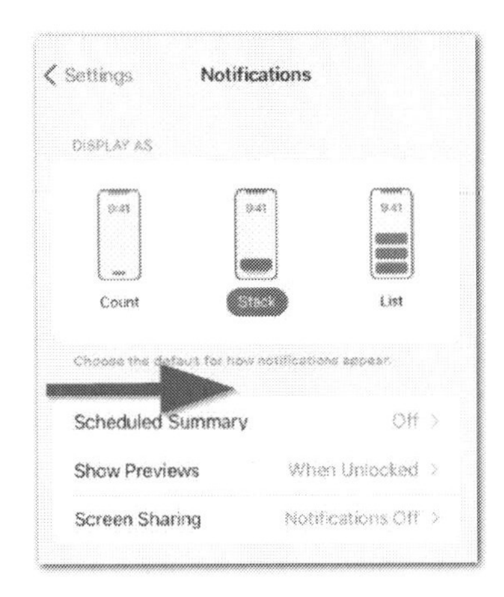

How to adjust the volume

1. You can use either **volume buttons** of your iPhone to adjust the volume when playing music, phone speaker, ringtone, and notification volumes.

Alternatively, use the **volume slider** in the Control Center as mentioned in the topic above or ask Siri to djust the volume, say, "Hey Siri, turn the volume up/down." Learn more about Siri in the next chapter.

ut iPhone in silent mode

1. Switch on **Ring/Silent,** so it turns orange, and put your iPhone in silent mode. Set the switch back to turn off silent mode, so your phone rings.

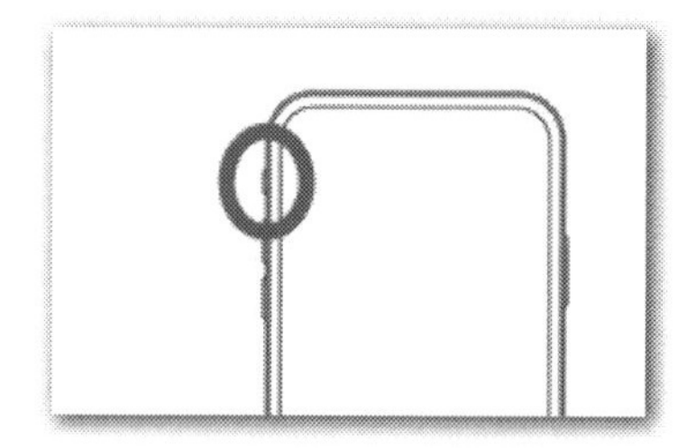

Note: The iPhone plays all sounds while the silent mode is off. It does not ring, play notifications, or make any other sounds when silent mode is on, except for clock alarms, media apps, and game audio.

How to Get to the Home Screen

1. iPhone SE's **Home button** is used to get home.

2. A simple way to get home is a quick swipe up from the white or black bar on the bottom of the screen called the **Home bar**. All the latest iPhone models have this kind of feature.

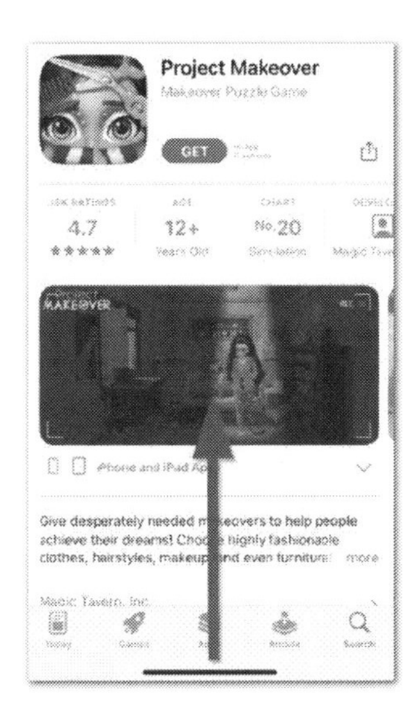

3. Alternatively, you can use AssistiveTouch to get to the Home screen. **AssistiveTouch** is a moveable button that you can use as a shortcut to the notification center, control center, Siri, device navigation, etc. This feature lets you do tasks by just tapping buttons instead of using gestures or pushing the side buttons, such as lock screen, adjust the volumes, summon Siri, and more. Tap the **AssistiveTouch button** to open the menu, then tap anywhere outside to close it.

'o get back to the home screen, tap the **AssistiveTouch button**, then tap the **Home button**.

CHAPTER 04

SIRI

What is Siri and How to Use It

Siri is a virtual assistant feature on iPhone that allows you to complete daily tasks with your voice. You can ask Siri to get a weather report, set an alarm, call a contact, play music, and many more. You can also set your preferred settings for how you want to make the most of Siri. This voice command function is available in any Apple product and is perfect for people who are busy completing tasks on their iPhone faster and more efficiently.

Because Siri is smart and interactive, it can answer questions and look up information from the web, use different language to command Siri, as well as interact with other apps to do actions for you, like posting on apps like Facebook and Twitter

Set up Siri

1. Go to **Settings**.

2. Scroll down or type **Siri & Search** on the search bar. Tap **Siri & Search**.

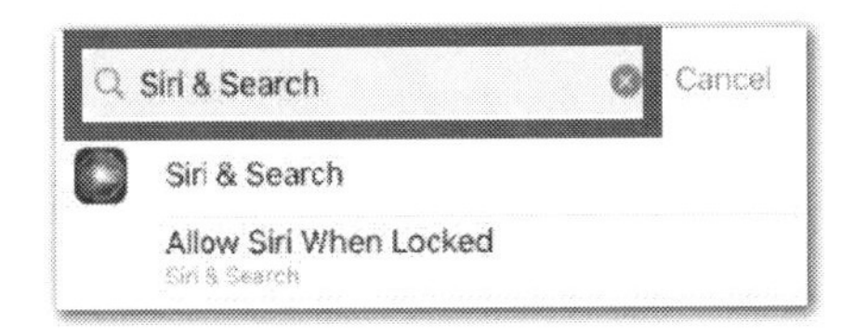

Enable Siri with your voice

1. Switch on **Listen for Hey Siri**, then tap **Enable Siri.**

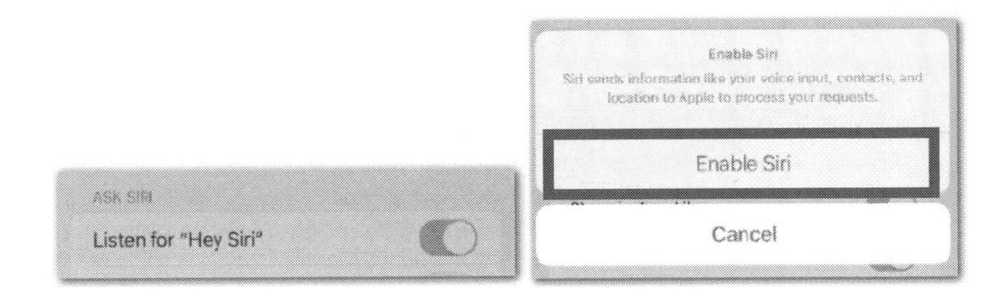

2. Tap **Continue.** Follow the prompt on the screen.

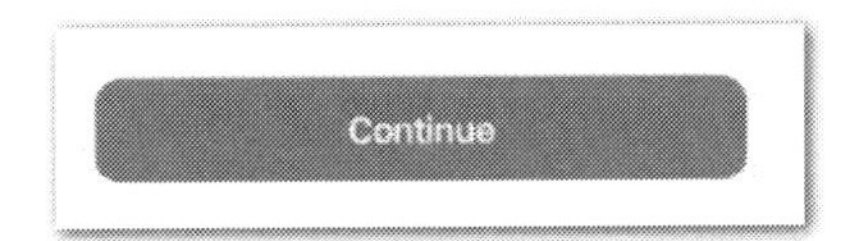

3. To make a request or ask a question, say "Hey Siri," followed by your request or question. Siri responds out loud to your command. For example, "Hey Siri, what time is it?"

4. To show what Siri says on the screen, tap **Siri Responses**. Switch on **Always Show Siri Captions**.

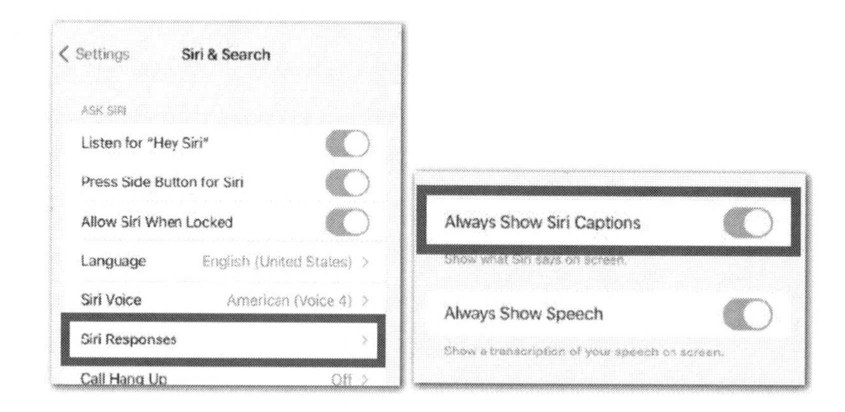

5. To show a transcription of your speech on screen, switch on **Always Show Speech**.

6. If you wish to disable Siri with your voice, switch on **Listen for Hey Siri**.

Enable Siri with a button

1. Switch on **Press Side Button for Siri** (for iPhones 13 & 14) or **Press Home for Siri** (for iPhone SE then tap **Enable Siri.**

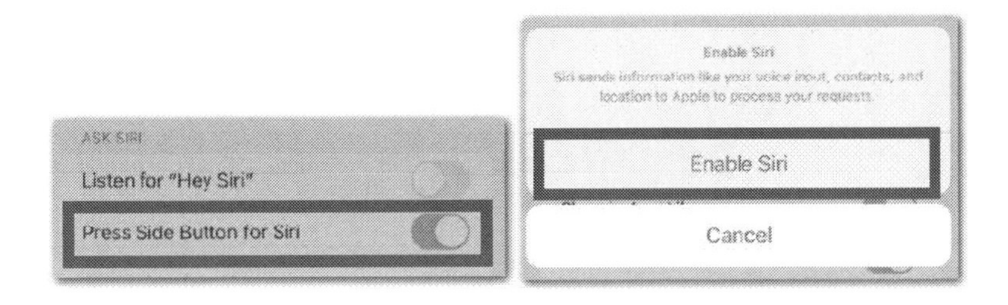

2. To make a request or ask a question, push and hold the **side button** (for iPhones 13 & 14) or push and hold the **Home button** (for iPhone SE), then ask a question or make a request.

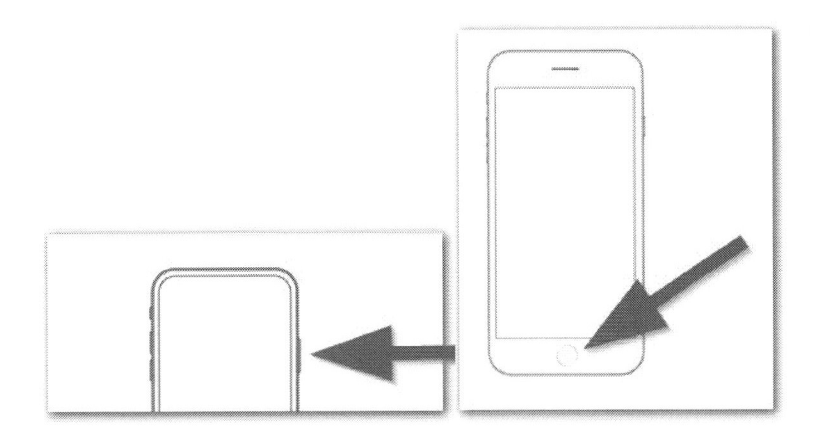

3. To disable Siri with a button, switch off **Press Side Button for Siri**.

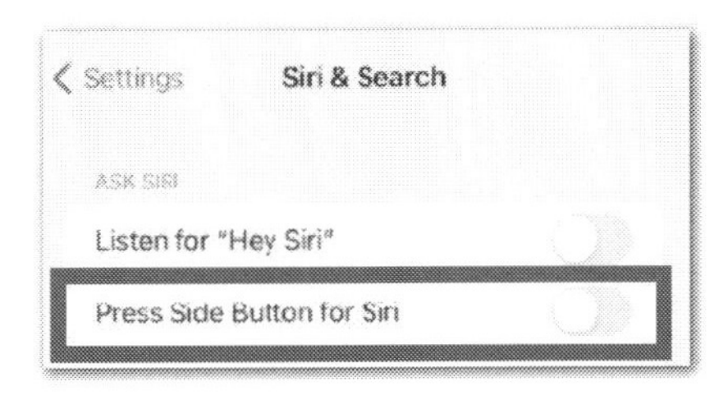

Make corrections

If Siri misunderstands your requests or questions, you can make corrections using the following ways.

1. Spell out your request. For example, "Call," J-O-H-N (spell out the name).
2. If you want to change your request, say, "Change it."

3. When you see your request onscreen, you can edit it. Tap the request, then use the onscreen keyboard to type a request.

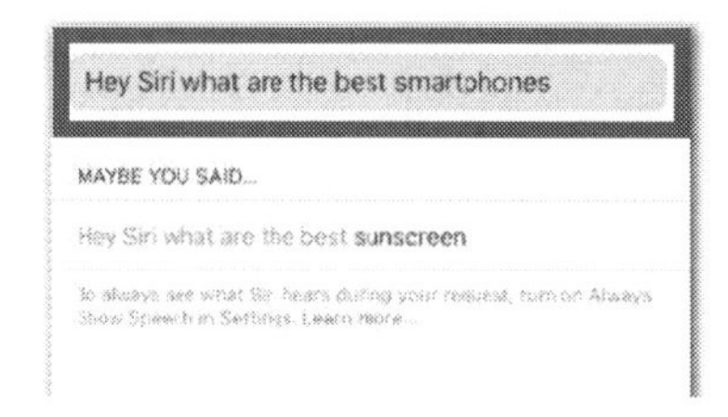

Set up type a request to Siri

4. Go to **Settings**.

5. Scroll down or type **Accessibility** on the search bar. Tap **Accessibility**.

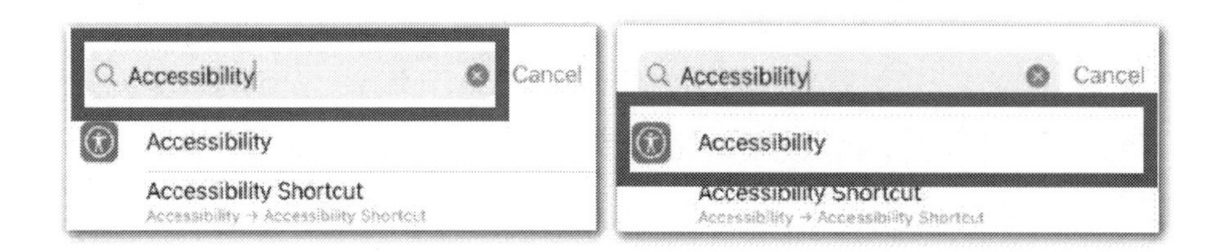

6. Tap **Siri** under General. Switch on **Type to Siri**.

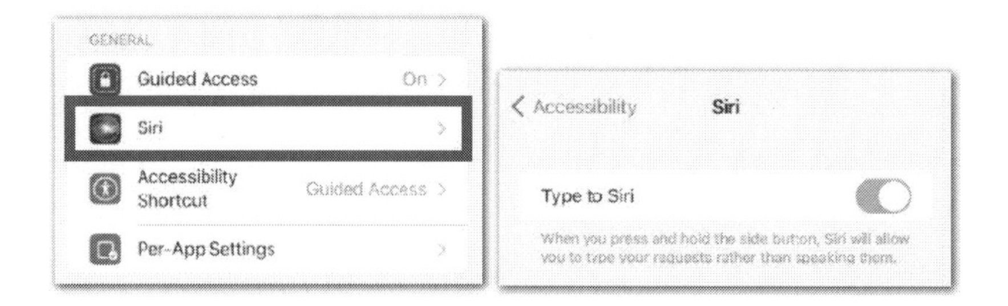

7. Type a request or question in the text field.

Ask Siri

Here are some of the questions you can ask Siri:

- "Hey Siri, what time is it?"
- "Hey Siri, how far is the closest pet shop from here?"
- "Hey Siri, what is 589,000 divided by four?"
- "Hey Siri, when is the best time to go to Australia?"
- "Hey Siri, how do you say, 'I left my wallet in Japanese'?"

Here are some tasks using the app that you can request from Siri:

- "Hey Siri, play Titanic on Spotify." Siri will play music on a music streaming app, Spotify.
- "Hey Siri, send a message to my wife saying, 'I'm waiting for you.'" Siri will send a message to a contact.
- "Hey Siri, directions to The Malt center." Siri will show results from a geo-location map app.
- "Hey Siri, show me Chinese restaurants." Siri looks up information from the web.
- "Hey Siri, add an appointment with Dr. Gomez tomorrow at 1 pm." This will be added as a reminder.

CHAPTER 05

COMMUNICATION

iPhone provides an excellent platform for communication with its users. With its many instant messaging options, including iMessage and FaceTime, users can stay in touch with friends and family without missing a beat, no matter where they are or what time of day.

Add Contacts to Your Phone

Add contacts to your iPhone in a few simple steps. Add more contact information for people you know, so you can send emails, use their mailing addresses, or add their birthdays to the reminder.

1. Go to **Phone**.

2. At the bottom of your screen, tap **Contacts**. Tap the **add button** at the top-right corner of your screen.

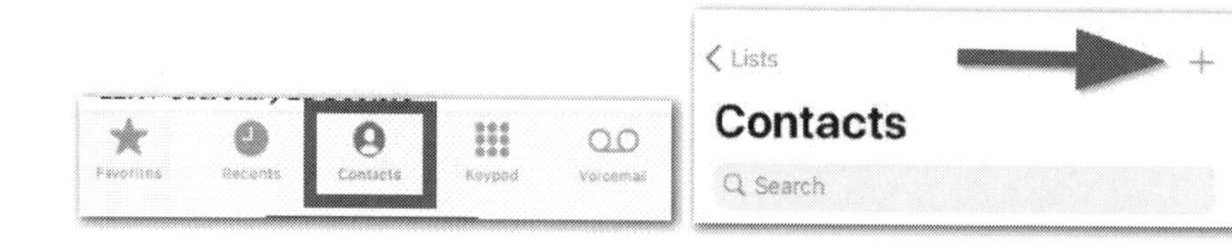

3. On the New Contact screen, input contact details.
4. You can add as much information as possible, such as another phone number, email, address, etc. To do this, tap the **add button** next to the label and enter the information required.

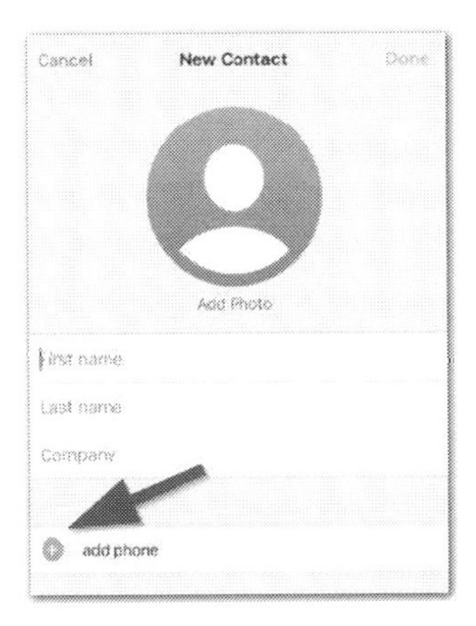

5. If you wish to remove the information, tap the **delete button** next to the label. Tap **Delete**.

·ind a contact

6. At the top of the contacts list, tap the search bar. Enter a name, phone number, or other contact information.

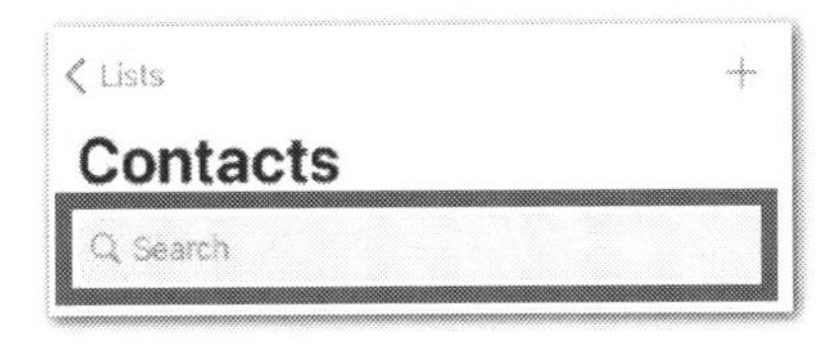

·reate a 'Favorite' Contact

ɔu may add anyone as your favorite contact.

1. Go to **Phone**.

2. Locate the contact you wish to put as a favorite, then tap **Add to favorites.** Select between the options where you want to add the contact as Favorite.

Alternatively, you can use the Favorite tab to add a contact as a favorite.

1. Go to **Phone**.

2. At the bottom of your screen, tap **Favorites**. Tap the **add button** at the top-left corner of your screen.

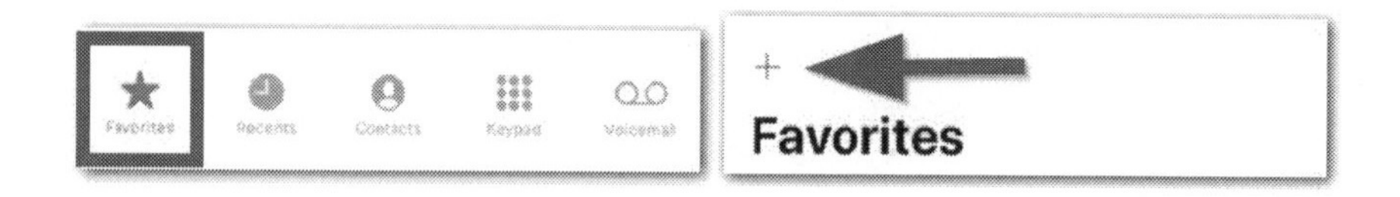

3. Select a contact to add as a Favorite, then select between the options where you want to add the contact as a Favorite.

Send and Receive Messages

Send and receive text, images, videos, and audio files with the Messages app. You can send a text message to one or more people to start a conversation.

1. Go to **Messages**.

2. Tap the **Create Message icon** at the top right corner of the screen to initiate a new message or respond to an existing one. Add the phone number, contact name, or Apple ID of your recipient, or tap **Add icon**, then select contacts.

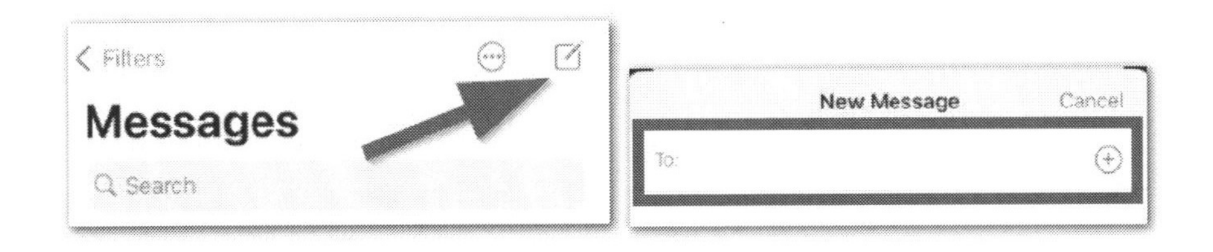

3. To send a message, type the message into the text field and tap the **Send icon.**

Send attachments

1. You may send texts, emojis, recorded audio, and attachments, such as Animojis, Memojis, images videos, and links, by tapping the **Apps icon** under the message field. Select which attachment wi be included in the message, then tap the **Send icon**.

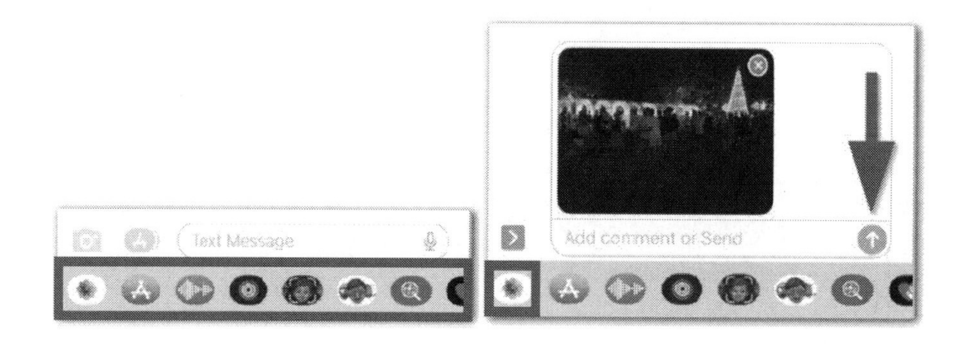

iMessage and Text Message

If the send button is blue, you'll be using **iMessage**, which is a messaging feature introduced by Apple an only works between iPhone and other Apple products; if it's green, your cellular service will be used to sen the **text message**. iMessages are sent over the internet or cellular data.

Switch messaging from iMessage to Text message

1. Go to the conversation.
2. Push and hold the message you want to send as a text message, then tap **Send as Text message**.

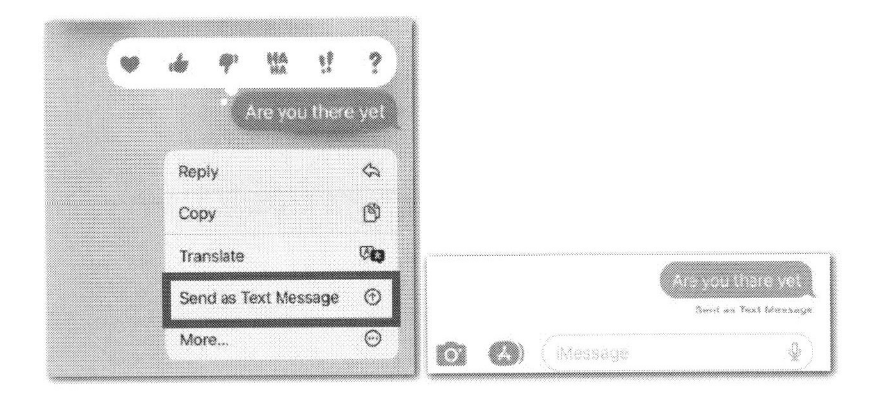

3. If a message can't be sent, an alert will show. Just tap the warning to send the message again.

eply to a message

conversation will pop up on your message list when a contact sends you a message. If you and this ərson have already had a conversation in Messages, their message will be appended to the end of the ıread. To reply to their message:

ا. Tap the conversation that you want to participate in. In the text field, type your message. Tap the **send icon** to send your message.

To let your recipients know you've read their messages, toggle the **Send Read Receipts** button:

1. Go to **Settings**.

2. Scroll down or type **Messages** on the search bar. Tap **Messages**.

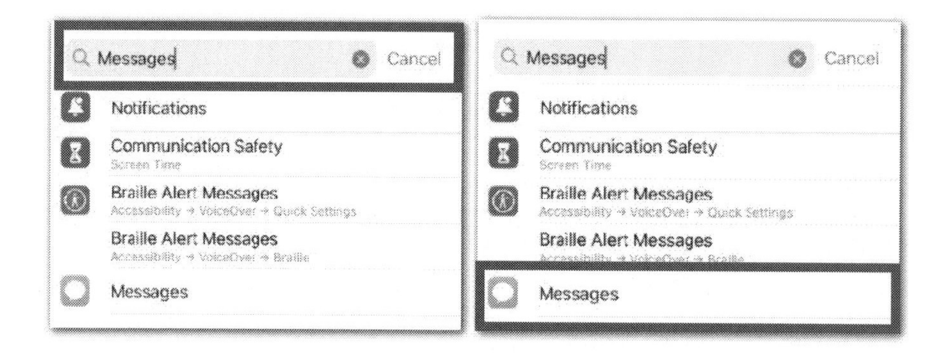

3. Scroll down to **Send Read Receipts** and switch it on.

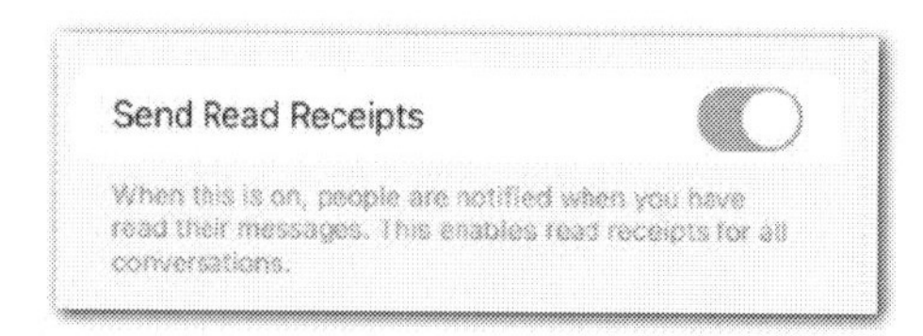

Create a Group Conversation

Send a mass text message to a group using the Messages app. You can get participants to pay attention to specific messages and work together on projects while chatting with a group. To start a new chat:

5. Tap **Create Message icon** from the menu bar.

6. Add the phone number, contact name, or Apple ID of each recipient, or tap **Add icon**, then select contacts.

7. In the text field, type your message, then tap **Send icon**.

\dd someone to an existing group conversation

’s possible to include more participants in a conversation if at least three people are already there (as long s everyone in the group uses an iPhone, iPad, iPod touch, or Mac and has turned on iMessage).

l. Tap the conversation you will add a new member to. On the top of the conversation, tap the **group name**.

2. Tap the group members' names, then tap **Add Contact** to add a new member to the group conversation.

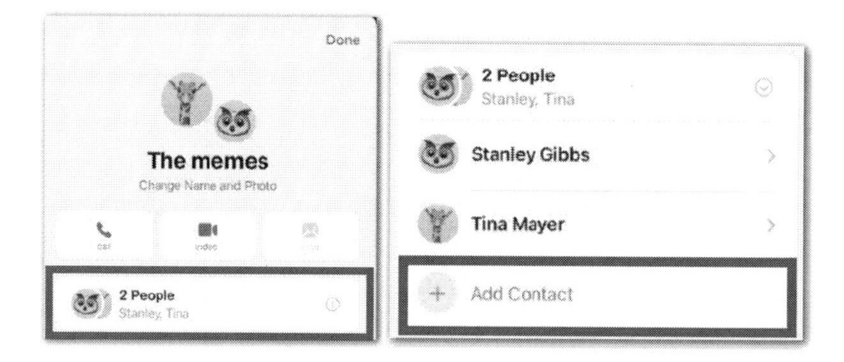

Note: Swipe someone from the discussion list by highlighting their name, then tapping Remove.

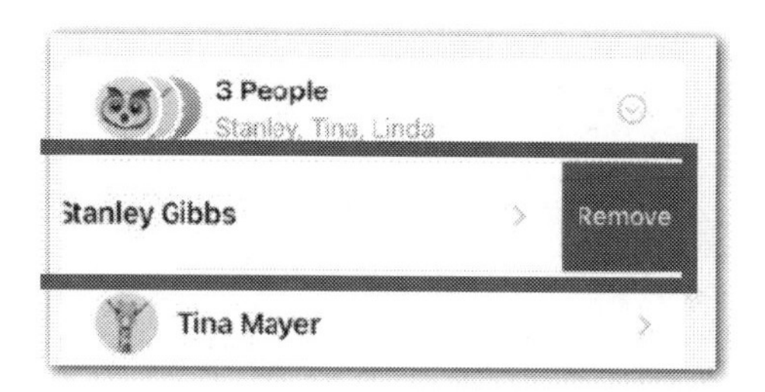

Leave a group conversation

A group message can be left (if everyone has an iPhone, iPad, iPod touch, or Mac and has turned on iMessage).

1. Select the group chat you wish to exit. At the top of the conversation, tap the group name.

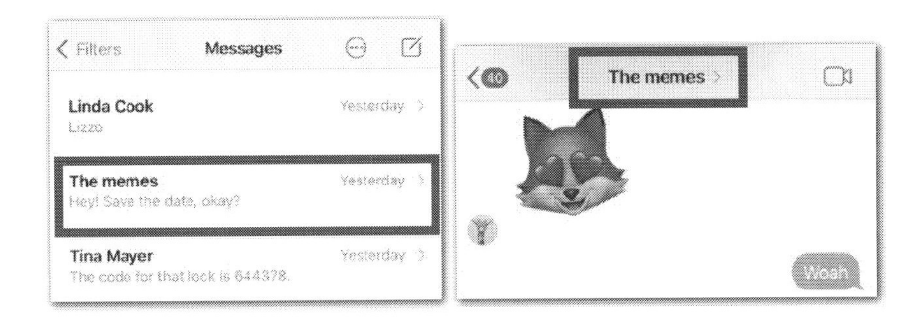

2. Scroll down and tap **Leave this Conversation**.

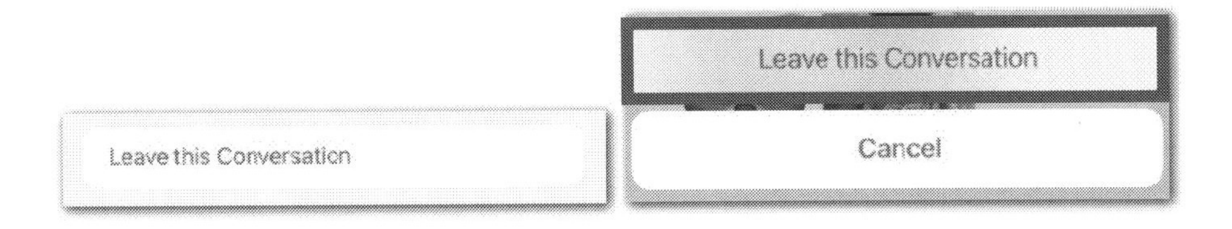

More Messaging Features

Forwarding Text Messages

1. Tap and hold the message bubble you want to forward, then tap **More**. You can also select other text messages that you want to forward, then tap the **Forward icon**.

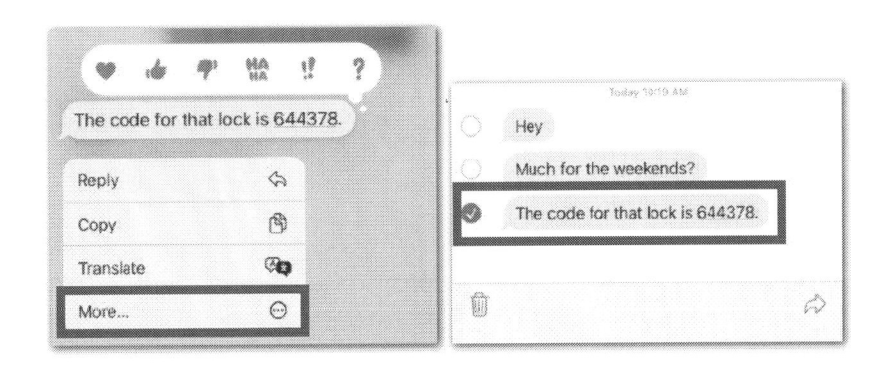

2. Enter a recipient, then tap **Send icon**.

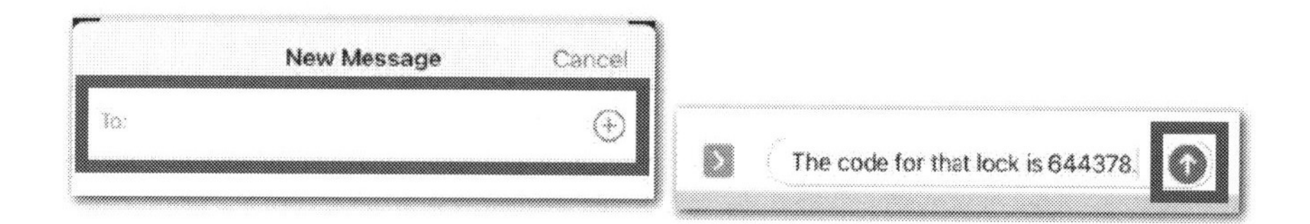

Delete and Mute Messages

3. Tap the message or conversation you want to mute or delete. Slide the message or conversatior the right, then tap the **Mute icon** to mute and the **Delete icon** to delete.

Copy and Paste Messages

1. To copy the message, push and hold the message bubble and select **Copy**. To paste the copi message, push and hold the text field and tap **Paste.**

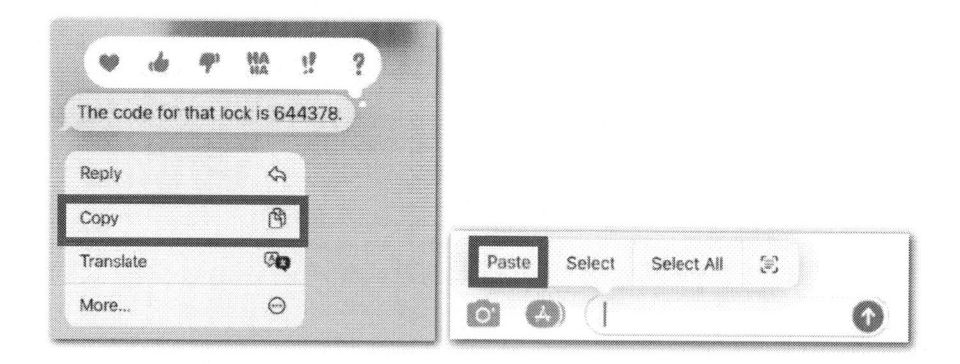

Making a Phone Call

You can make a call in the Phone app by entering the number manually, selecting an already made or saved call, or selecting a contact from your address book.

Dialing a number in the Phone app

1. Go to **Phone**.

2. Select **Keypad,** then enter the number. Tap the **delete icon** in the right bottom area to clear input.

3. For international calls, a "**+**" should be pressed before adding the number. Push and hold the "**0**" key until "**+**" shows.

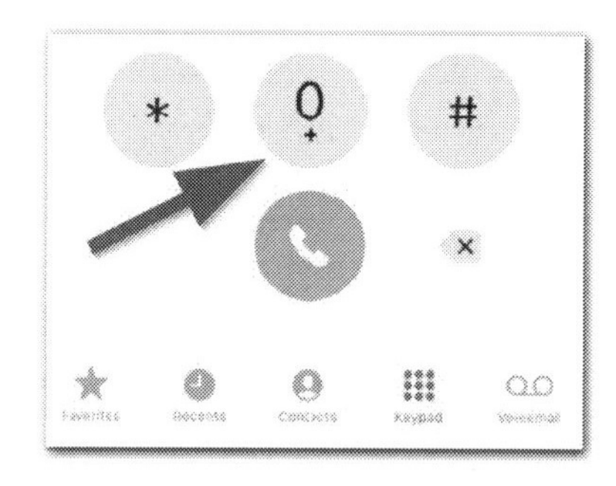

4. Tap the **Phone icon** to start the call. To end the call, tap the **End icon**. You may also place a ca directly from any website or messaging application by tapping the number and selecting call.

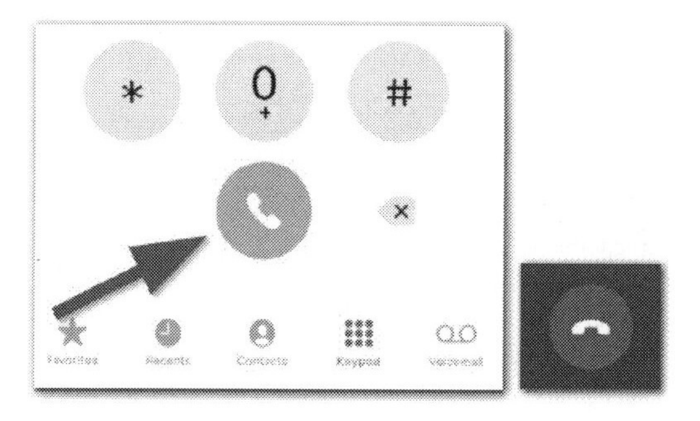

Calling someone from a contact list

1. Go to **Phone**.

2. Tap **Contacts**. Select the contact, then choose the phone number you want to call.

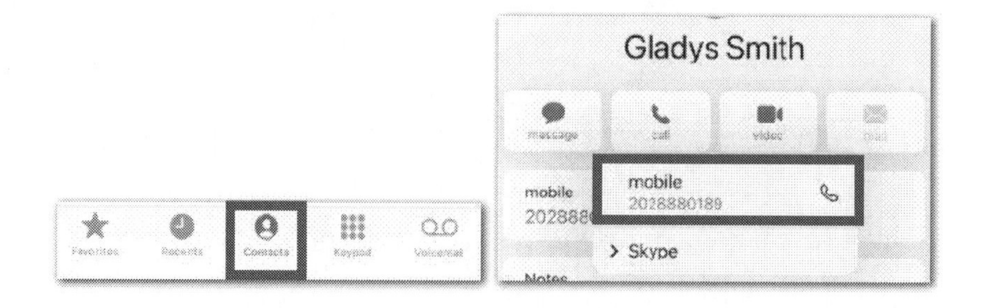

Answer or decline incoming calls on iPhone

You have the option to either accept an incoming call or ignore it. A call is sent to voicemail if you decline it. You can opt to respond with a text message or call back.

Answer a call

1. Tap the **phone icon**. If iPhone is locked, drag the slider.

Silence a call

1. Push the **side button** or volume **button**. A silenced call can be answered until it goes to voicemail.

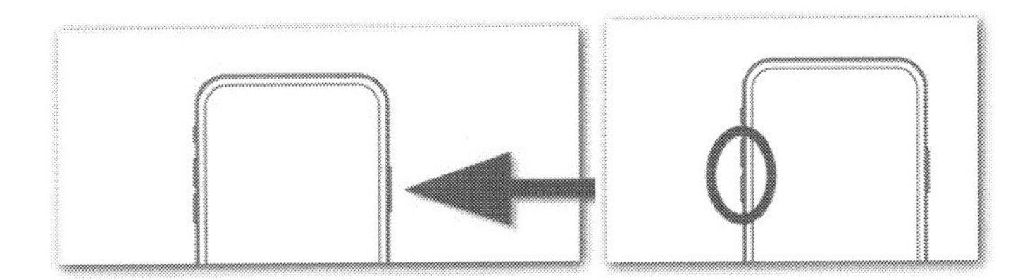

Decline a call and send it directly to voicemail

1. Quickly push the **side** button twice or tap the **decline or end icon**.

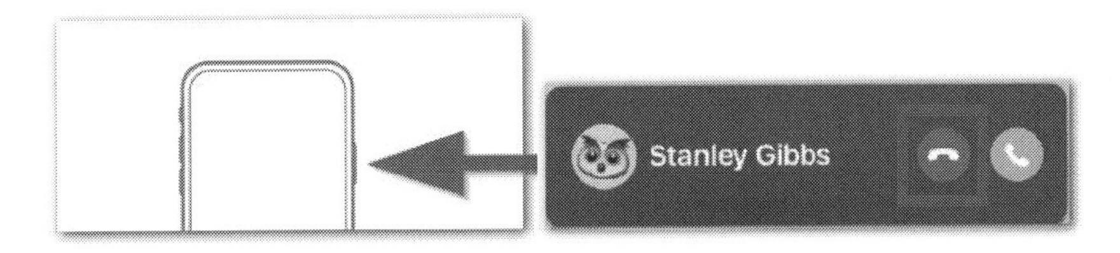

Audio call adjustments

The iPhone's call volume can be adjusted via the volume buttons. You can also accomplish any of the following by swiping down on the call banner:

1. To mute: Tap the **mute icon**.

2. To put the call on hold: Push and hold the **mute icon**.

3. To talk hands-free: Tap the **audio icon**, then select an audio destination.

4. To use the keypad: Tap the **keypad icon**, then enter digits.

5. To access contacts: Tap the **contacts** icon, then select a contact.

6. To add another call: Tap the **add** icon and select a contact you want to add to the call, then tap the **merge** icon.

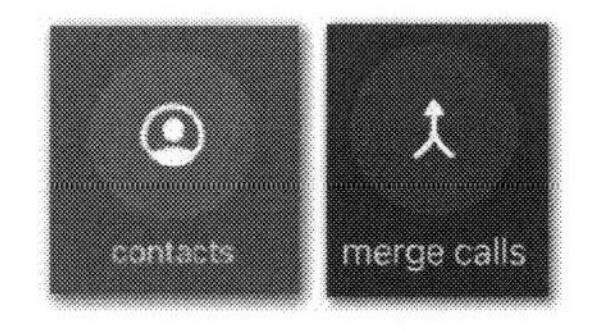

Phone Voicemail

rom the Phone app, you may check your Voicemail (offered by some service providers). Without listening ɔ every single one, you have the option of selecting which ones to play and which ones to delete. The umber of unread messages is represented by a badge on the Voicemail icon.

et up Voicemail

he first time you use Voicemail, you'll be prompted to set up a voicemail password and record a personalized reeting.

. On the Phone screen, tap **Voicemail**, then tap **Set Up Now**.

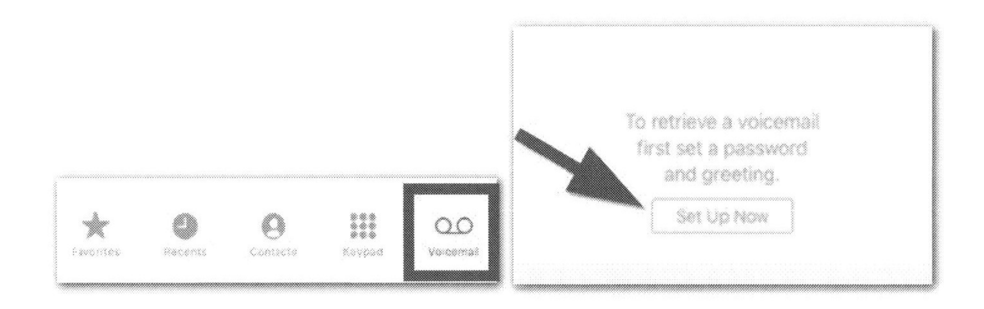

2. Enter a secure voicemail passcode, then tap **Done** at the top-right corner of your screen. Select a greeting, either **Default** or **Custom**.

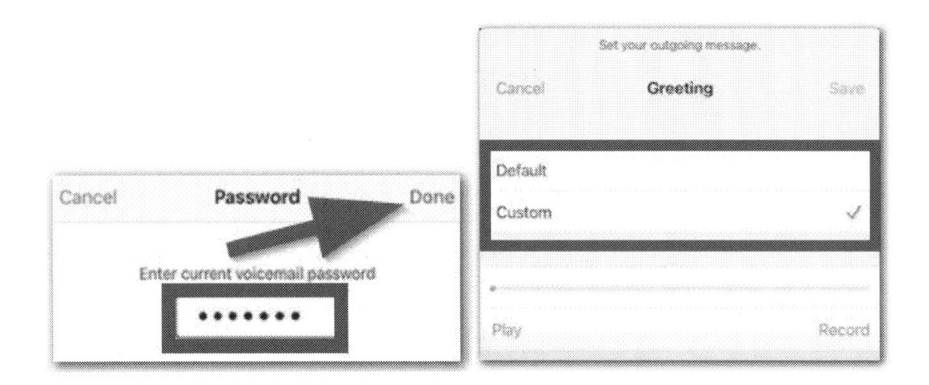

3. With Custom, you can record your greeting. Tap **Record** to record a voice recording. Tap **Play** to listen to it.

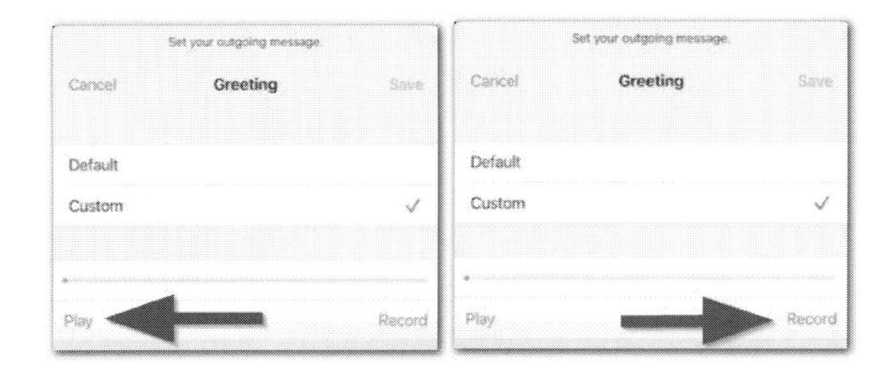

4. 4. Tap **Save** at the top-right corner of your screen.

Play, Share, or Delete a Voicemail

1. On the **Phone** screen, tap **Voicemail**. Tap a message.

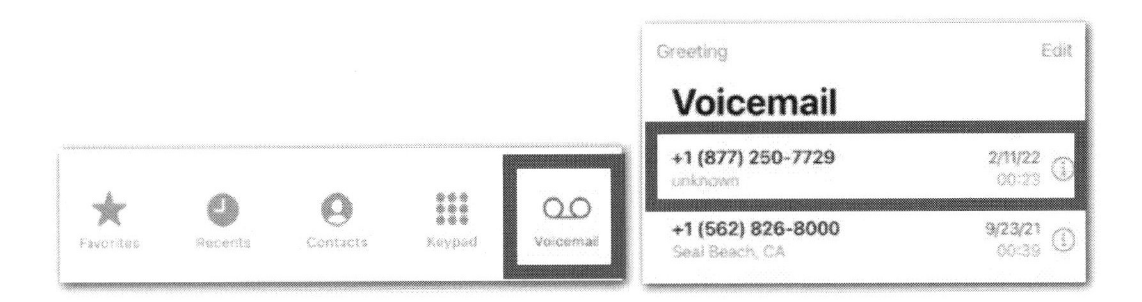

2. To play the message: Tap the **Play** icon.

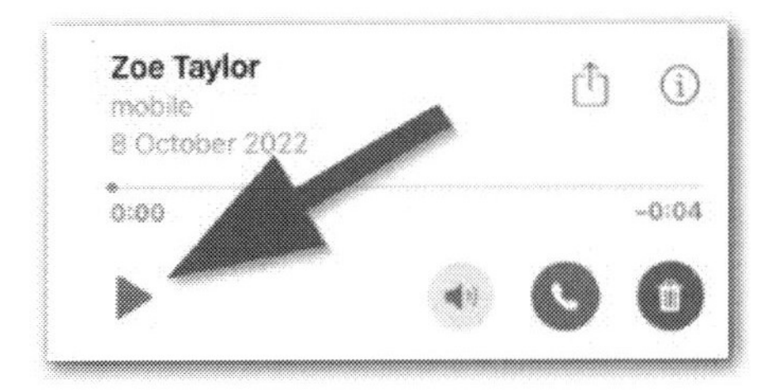

3. To share the message: Tap the **Share** icon.

4. To delete the message: Tap the **Delete** icon.

ote: Your carrier may permanently delete messages in some countries or regions. Changing your SIM ard could also delete your voicemails.

To recover deleted messages, tap **Deleted messages**. Select the message you wish to restore and th
tap **Undelete**.

How to Use Dictation

iPhone has a speech-to-text conversion feature called Dictation that transcribes speech into text. Th
functionality captures and converts your voice message to the text displayed on the phone's screen. Y
can check the transcribed text for accuracy and edit it later. Dictation is designed to assist users who ne
faster typing or who want relief from writing long messages. Using Dictation is easy and does not requ
an internet connection.

Tip: Speak slower for better speech recognition and try to speak louder and more clearly to avo
misunderstandings.

Dictate a text

1. Tap the **microphone** icon on the text field or keyboard, then start speaking. You can see the text o
the screen as you talk. Stop speaking when done, then tap the **microphone** icon on the keyboard
disable dictation.

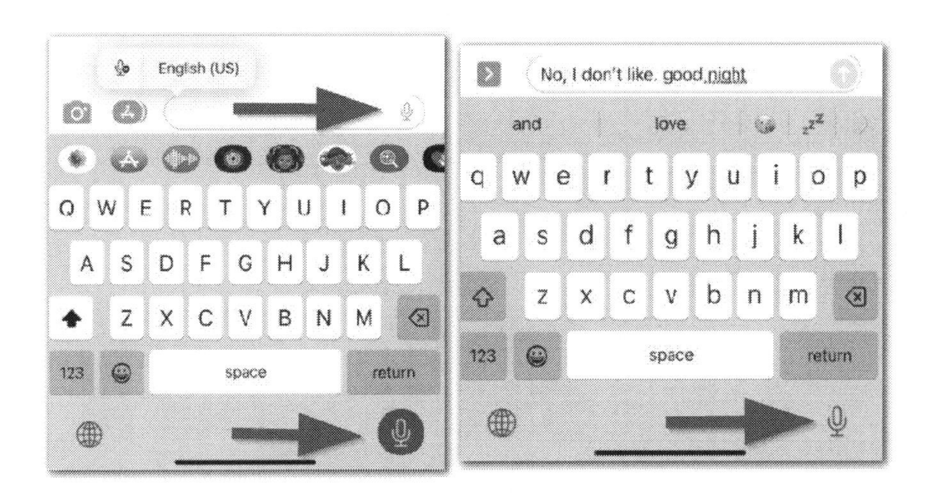

2. When you see a blue line underlining the text, your speech is unclear for Dictation. Tap the **underlined words**, then type the right words to correct them.

3. You can also dictate to make corrections. Double tap the **underlined words**, tap the **microphone icon**, and speak again to replace the word. Tap the **microphone icon** on the keyboard to finish. Select all the texts you want to replace, so you can correct them.
4. If you wish to change the language in Dictation, push and hold the **microphone button**, then select another language.

Punctuation and formatting

Dictation can automatically insert punctuations in your texts. Use punctuation in your voice command like "Are you coming question mark," which becomes "Are you coming?" Below are some common punctuation and formatting terms you can use in Dictation:

- Period: .
- Comma: ,
- Colon: :
- Equal sign: =
- Dollar sign: $
- Percent sign: %
- Trademark sign: ™
- Copyright sign: ©
- Ampersand: &
- Smiley: :-)
- Open parenthesis: (
- Close parenthesis:)
- New line: creates a new line
- New paragraph: creates a new paragraph
- Space bar: space

What is FaceTime and How to Use It

FaceTime is an easy-to-use video messaging app that allows you to communicate with your friends and family, whom you can see, listen and chat with on your device. It's done over cellular data or Wi-Fi and works between Apple products like iPhone, iPad, and Mac. There are two things you need to use this video call: a phone number of your contact or an email address. Enjoy a two-person call or a group FaceTime that accommodates more than 20 people in a video call room that appears in small tiles on the screen.

Make a FaceTime call.

1. Go to **FaceTime**.

2. Tap **New FaceTime**, then type a contact name, phone number, or email address.

3. Tap the **call button** or **FaceTime icon**. To stop the FaceTime call, tap the **close button**.

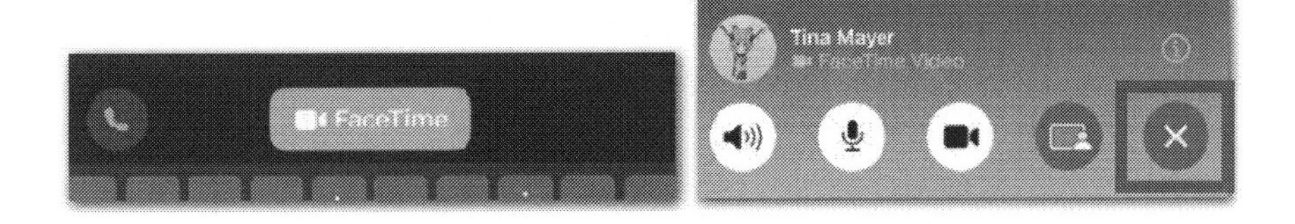

Jse FaceTime on a phone call

4. You can also make a FaceTime call directly in Contacts. Tap a contact's name, then tap the **FaceTime icon**. While on a call, you can use your iPhone to initiate a FaceTime video call. On the Phone screen, tap the **FaceTime icon**.

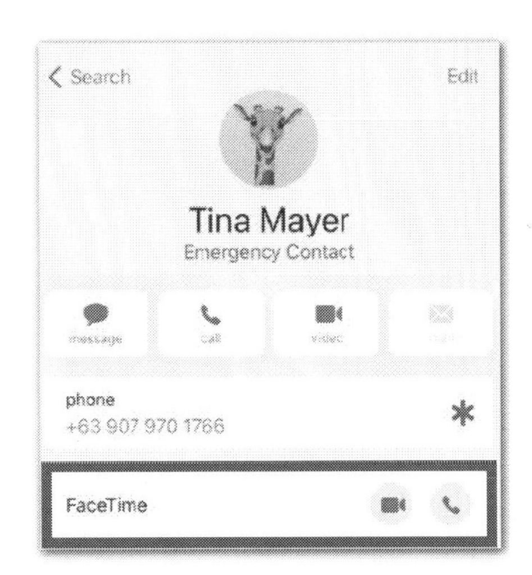

Answer a FaceTime call

1. When you receive an incoming FaceTime call, you can answer it from the call notification that shows anywhere on the screen. Tap the **call icon** or the **FaceTime icon** to answer it.

Create a link to a FaceTime call

Create and send a FaceTime call link to others via text message, chat apps, or email. The recipients car use the link to join your FaceTime call when received.

1. On the **FaceTime** screen, tap **Create Link**. Tap **Add Name**.

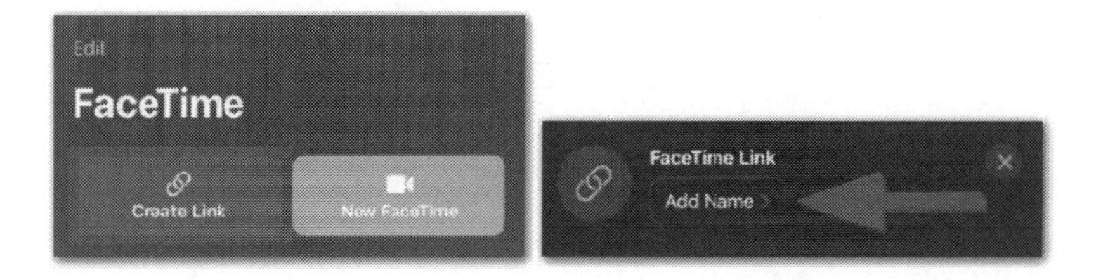

2. Type a meeting name, then tap **OK**. Tap the **close icon**. Send the link to the suggested contacts c send it via Messages, Mail, or any other supported apps in the Share menu.

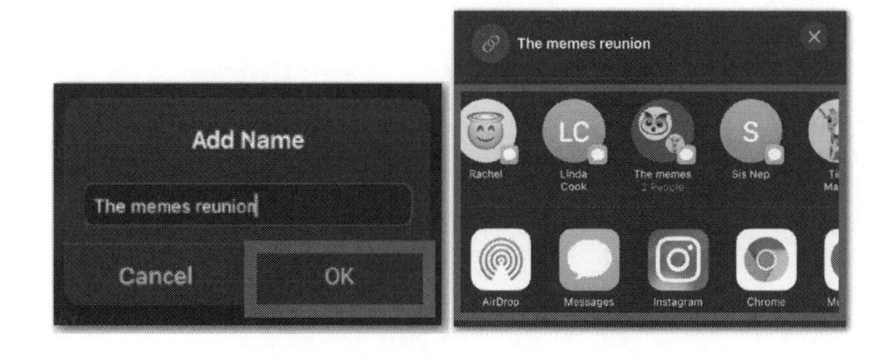

3. To join using the FaceTime link, tap the link on the FaceTime screen. Tap **Join** and wait for the other participants to get in.

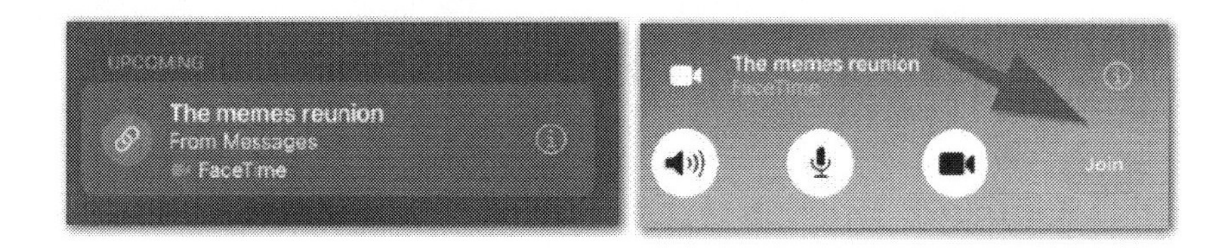

FaceTime is unavailable to non-Apple users, but they can join a video call set up from an Apple device. They can tap a FaceTime link and get on the call via web browsers.

Share screen in a FaceTime call

You can share your screen during a FaceTime call with everyone in the conversation. It means that what you see on your phone screen is what others can see. Participants are still visible in small tiles and can respond while you share your screen. To start sharing your screen:

1. On a group FaceTime, tap the screen to show the controls, then tap the **screen icon**. Tap **Share My Screen**. A countdown will start on the screen icon, after which your screen will be shared in the group FaceTime call. All participants of the group call will be able to see your screen.

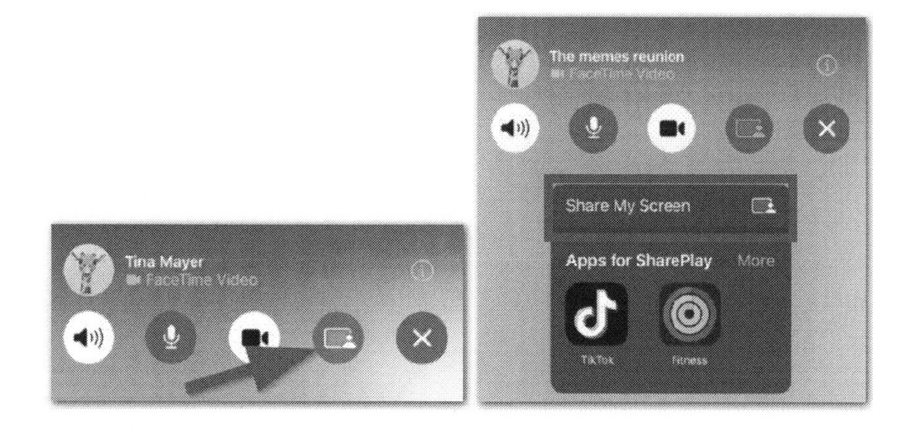

2. You can share whichever on your phone screen you want to share, such as an image, webpage, or video. To stop screen sharing, tap the **screen icon** once.

How to Block Contacts

Knowing how to block your contacts is good to know when there are unwanted phone or FaceTime calls and messages from unknown people you want to avoid. In today's technology, many scammers can lurk behind unidentified mobile numbers to take advantage of ordinary phone users. It's essential to be cautious when receiving communication, such as calls from a person asking for your personal information or messages that contain suspicious content, such as clicking a link. If it seems fishy, you can block a contact for your safety.

1. Go to **Phone**.

2. Search the contact you want to block from the Favorites, Recents, Contacts, or Voicemail tab. Tap the **info icon** next to the contact. Scroll down, then tap **Block this Caller**.

CHAPTER 06

CAMERA & MEDIA GALLERY

Undoubtedly, one of the key details people enjoy about the iPhone is its camera quality. Apple takes pride in its camera app's powerful snapshot capability, featuring several camera and video modes, lighting, and motion features. With the proper techniques and mastery, users can upgrade their mobile photography and video recording to a professional level. Plus, there are many more exciting ways of viewing and editing your photos with the media gallery.

How to Take a Photo

Play with your camera by using different modes and exciting camera functionalities. For best photo results, hold your camera still to avoid blurry photos. If you want to turn off the shutter sound when taking a picture, switch **Ring/Silent** on, so it turns orange to put your iPhone in silent mode.

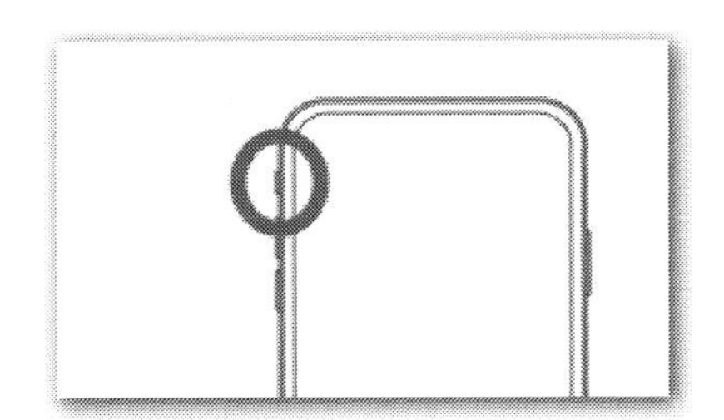

1. To go to the Camera, tap the **Camera** app on the Home screen.

Alternatively, you can tap the **Camera icon** on the Lock screen or swipe left from the Lock screen.

2. Tap the **Shutter button** or hold either volume button to take a photo. To switch between the bac and front camera, tap the **switch icon** at the bottom-left of the screen.

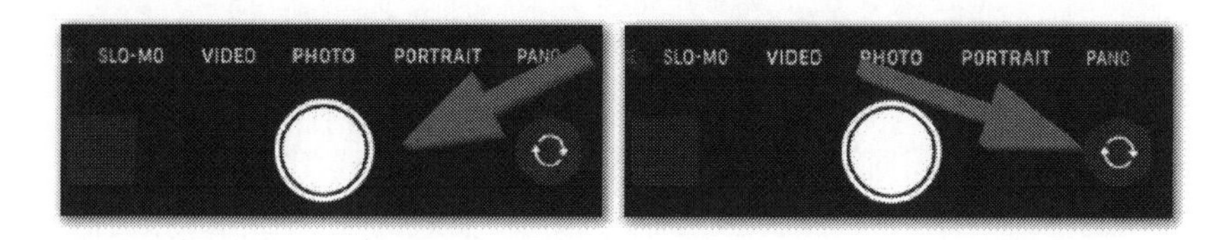

Use flash and night mode

1. To turn on a camera flash, tap the **flash icon** on the top-left corner of the camera screen.

2. Night mode is on by default. When it detects a low-light environment, it gives you a signal to hold still, so you can capture better photos. Tap the **night mode icon** to turn it off, so it will be slashed.

Zoom in and out

1. Tap **0.5x** or **1x** at the top of the shutter button to zoom out and zoom in, respectively or pinch the screen.

ake a Live photo

Live Photo is a camera feature in iPhone that captures the scene before and after you tap the shutter. ve photo records for 1.5 seconds only and is activated by default.

. Tap the **shutter button** to take a Live photo. To turn it off, tap the **Live photo icon** at the top-right of the screen, so it will be slashed.

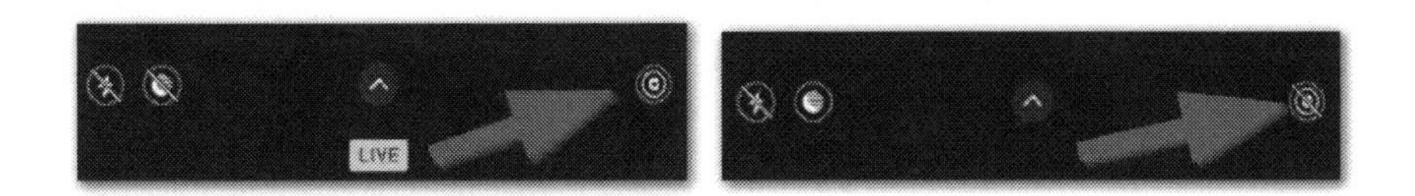

iPhone Camera modes

Swipe left or right on the screen to switch between camera modes.

A. **Photo** – the default camera mode when you open Camera. This is used for standard shots of still and live photos.

B. **Pano** – grabs a panorama picture that shows a complete view or wider angle in a single image. To capture using Pano, move the iPhone and follow the arrow on the screen. When you're happy with the image, you can finish by tapping the **shutter button** even before the arrow meets the end of the screen

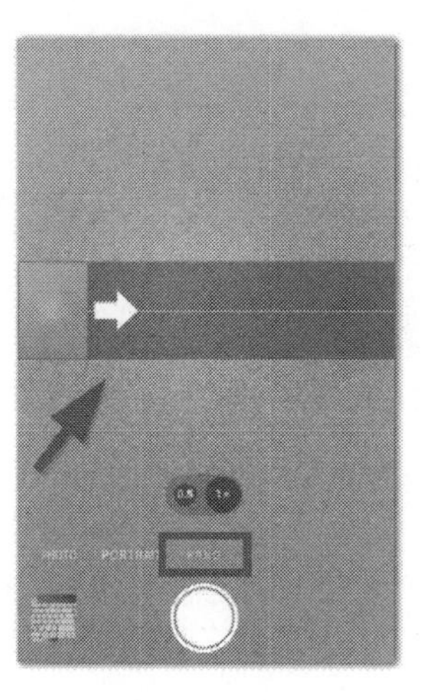

C. **Portrait** – focuses beautifully on the subject when taking a photo. Select your desired effect from th available options, then tap the **shutter button** to capture.

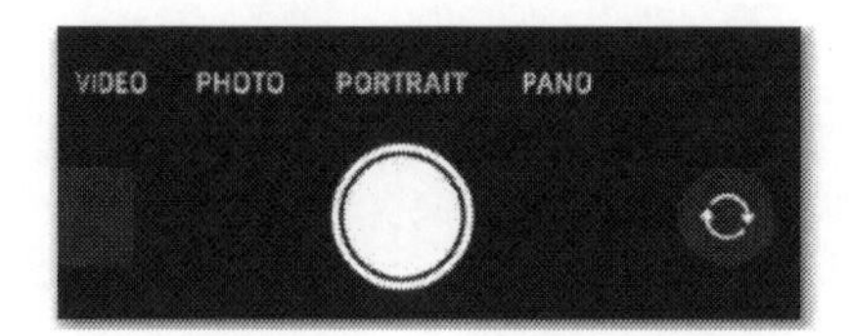

How to Record a Video

Using the Camera app, you can record a video. The video screen has a simple look, featuring the back and front camera switch, flash, time recorder, and screen sharpness settings.

1. Tap the **Video recording button** to start and stop recording. To switch between the back and front cameras, tap the **switch icon** at the bottom-left of the screen.

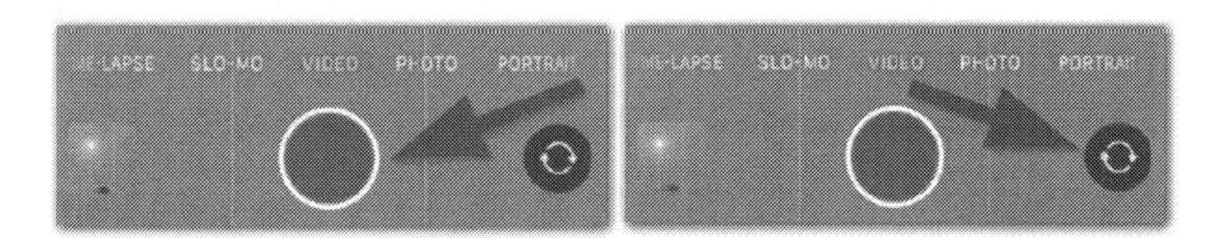

Use flash

2. To turn on a camera flash, tap the **flash icon** on the top-left corner of the camera screen.

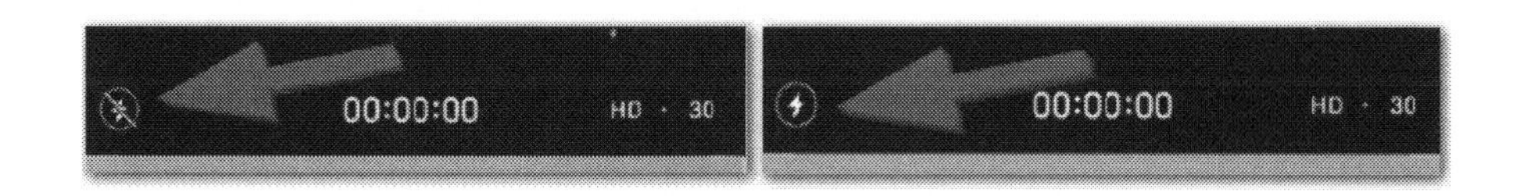

oom in and out

Tap **0.5x** or **1x** at the top of the shutter button to zoom out and zoom in, respectively or pinch the screen.

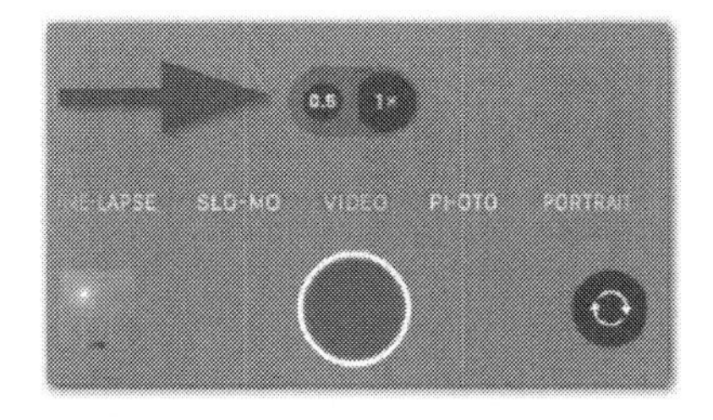

4K and HD

1. Tap **4K** or **HD** at the top-right of your screen to switch between the two. 4K and HD are screen settings that make your screen display sharper. 4K gives a sharper display than HD.

Video modes

Swipe left or right on the screen to switch between video modes.

A. **Video** – records a standard video.

B. **Time-lapse** – creates a video shown in a quick shift after it was recorded for a long period. Fo example, a video of the sun rising until the sun sets is recorded in 10 seconds.

C. **Slo-mo** – records a video in slow-motion.

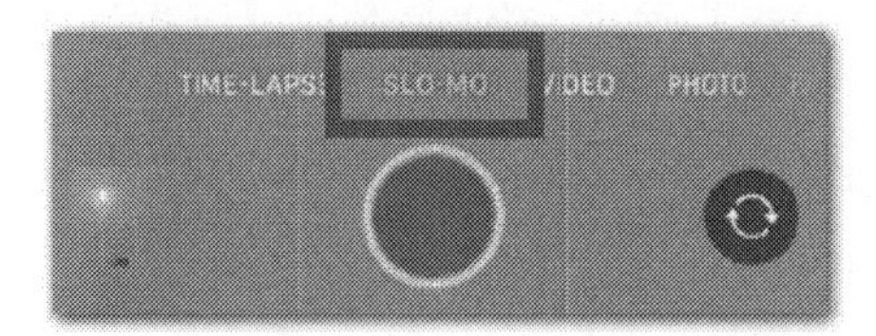

D. **Cinematic** – applies a cinema-grade effect to your videos. iPhones 13 and 14 support Cinematic mode.

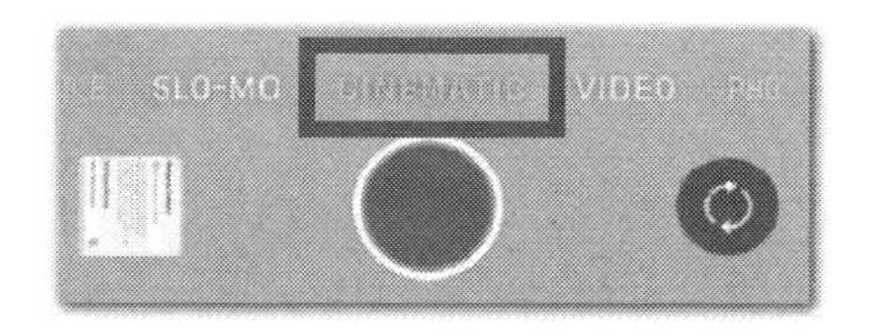

You can view recently taken photos and videos by tapping the small square on the bottom-left of your camera screen.

iPhone Gallery

Photos is the gallery of images and video recordings on your device. It is where all the media taken from your camera is stored. From here, you can view, play, and edit pictures and videos, as well as organize them. This topic will cover the basic features and navigation offered in Photos.

View photo gallery

You can choose how to browse your photo gallery using the following tabs at the bottom of your screen.

A. Library – Displays gallery by days, months, years, and all photos.

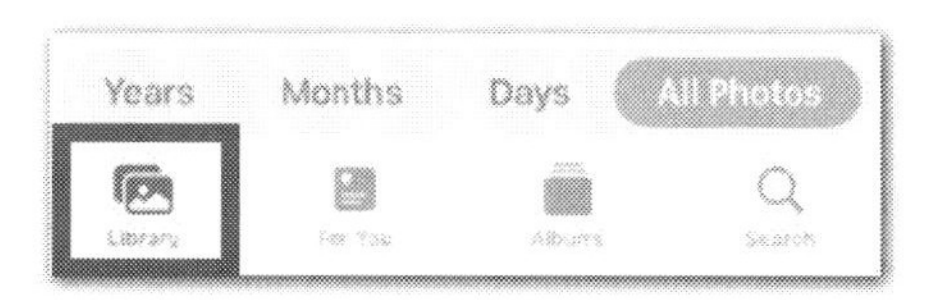

B. For You – Shows gallery in categories like memories, shared photos, featured photos, and sharing suggestions.

C. Albums – Displayed photos and videos in albums you create, favorites album, and other app categories where the images were captured. It also shows other collections according to people, places, media types, and utilities.

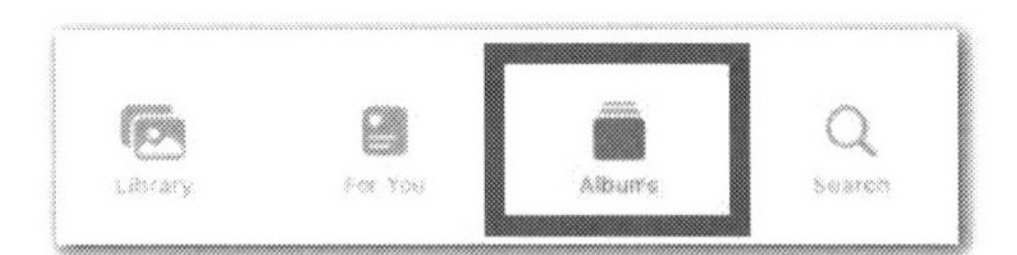

To create an album:

1. Tap the **add button** on the top-left corner of the screen, then tap **New Album**.

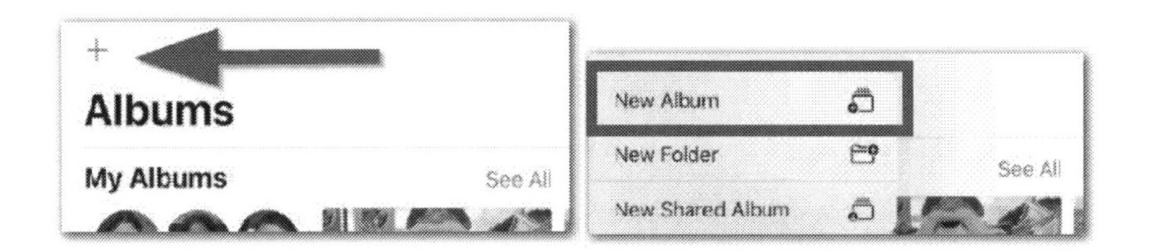

2. Enter a name for your album, then tap **Save**.

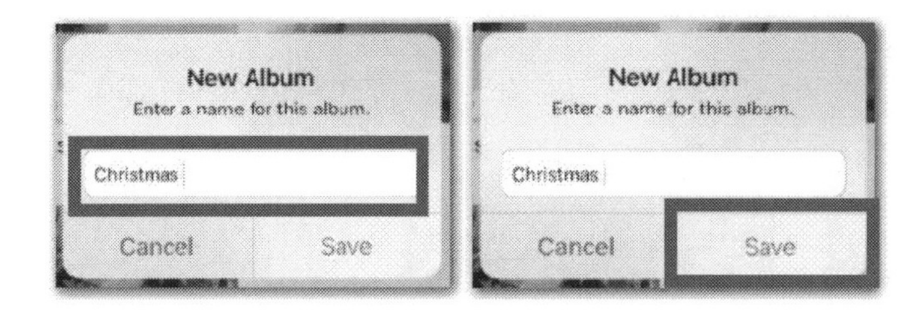

3. Select photos to add to your album, then tap **Add** at the top-right corner of your screen.

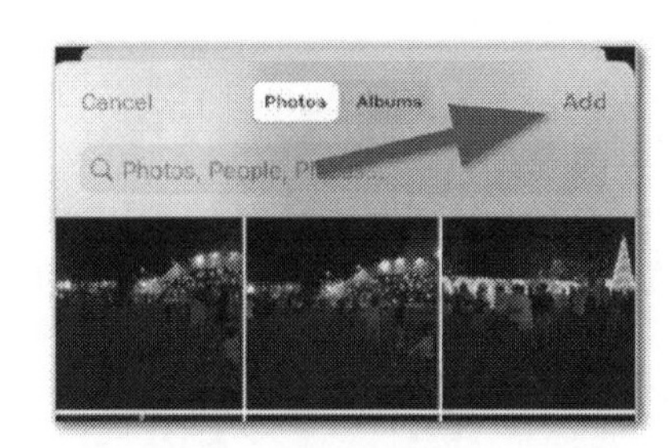

D. Search – This lets you search for your photos via keywords. Enter keywords, such as people, places, objects, or locations in the search field, to search for matching pictures.

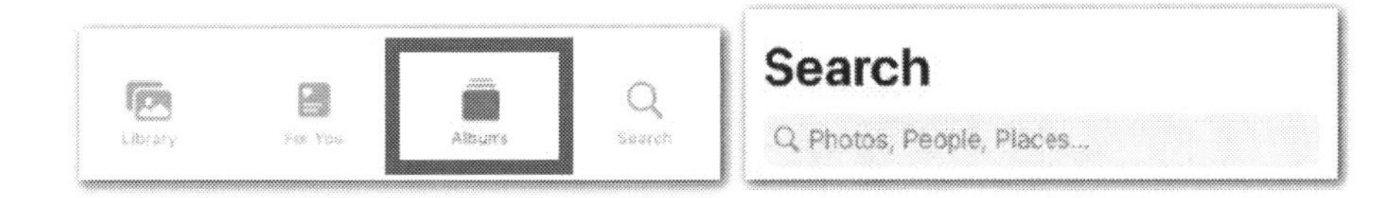

View a single photo

1. Tap a photo for a full view. Double tap to zoom in and out on the photo.

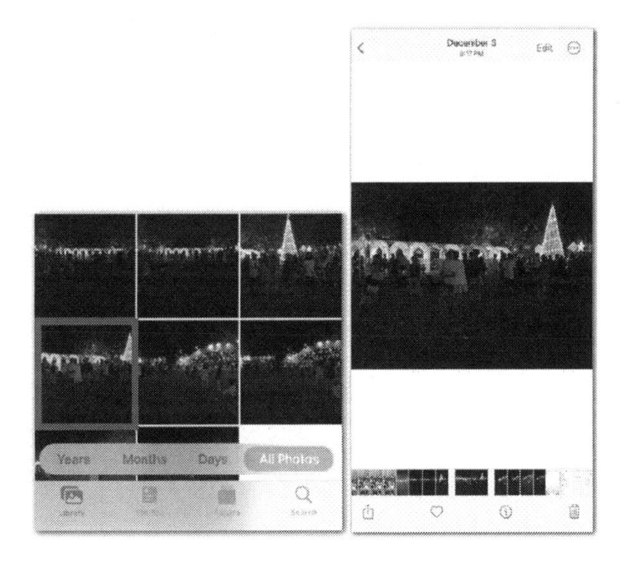

ore features when viewing a single photo:

reate actions on a photo

. Tap the **share icon** at the bottom of your screen to create many actions on the photo, like sharing it using apps, copying, adding to an album, and more.

Add a photo to Favorites

1. Tap the **heart icon** at the bottom of your screen to add a photo to the Favorites album.

View photo information

1. Tap the **information icon** to view details about the photo, like the date and time taken. You can also adjust the date and add locations with this menu.

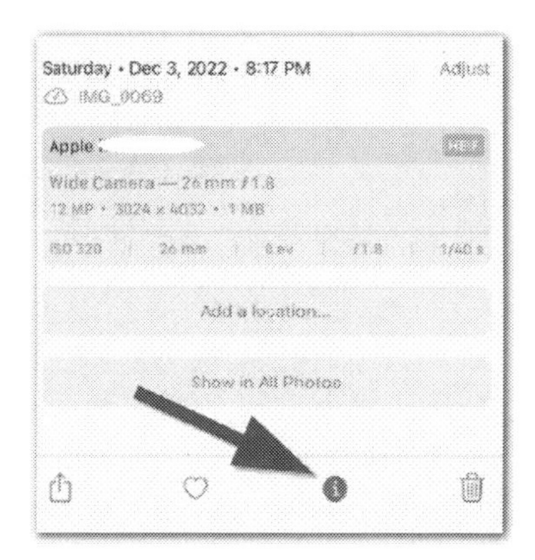

Delete a photo

1. Tap the **delete icon** at the bottom of your screen to remove the photo from the gallery. Tap **Delete Photo**. The deleted photo remains in the "Recently Deleted" folder for up to 40 days after the deletion. After that time, the photos will be permanently deleted.

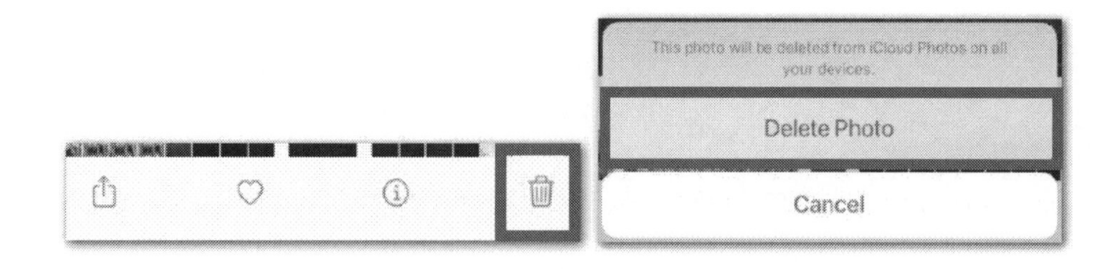

2. If you wish to restore a deleted photo, go to **Albums**, scroll down to Utilities, then tap **Recently Deleted**.

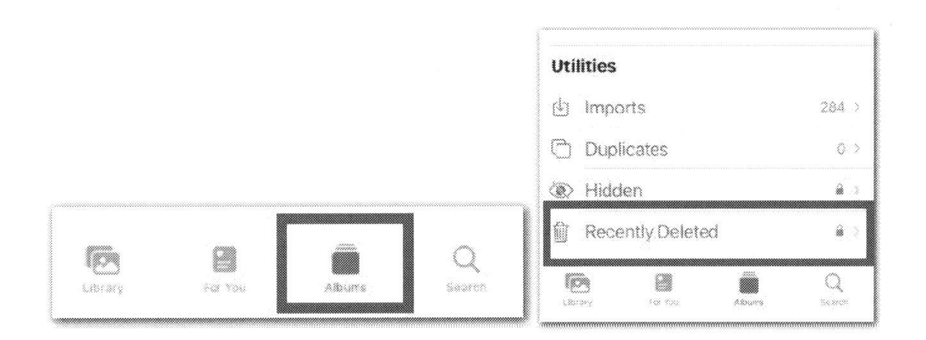

3. Tap the photo you wish to restore, then tap **Recover** at the bottom-right of the screen. Tap **Recover Photo**. The restored photo then returns to the gallery.

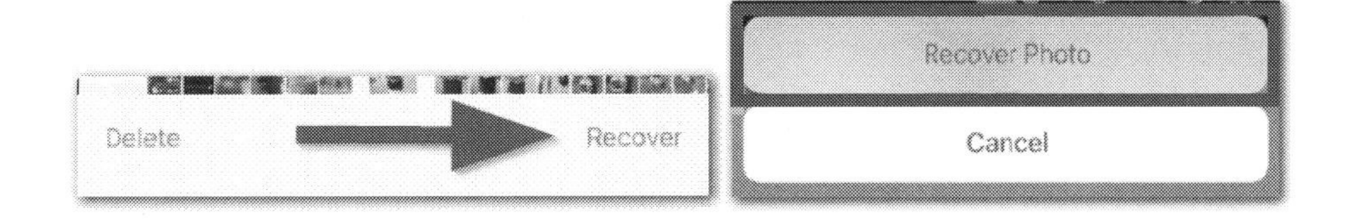

dit a Photo

. Tap on **Edit** at the right-left of the screen. At the bottom of your screen are the basic editing features you can apply to your photo. **Tip:** Tap **Auto** to automatically apply editing that fits your image, so you don't have to change other elements.

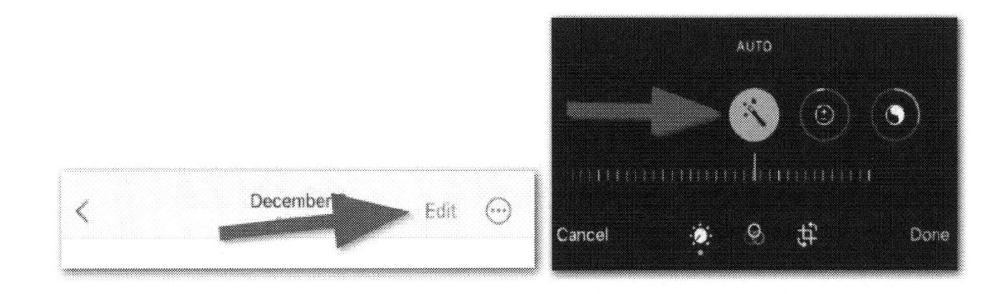

. You can also apply a filter to your photo. Tap the **three-circled icon** and swipe left and right through the filter gallery to select a filter. If happy, tap **Done** at the bottom-right of your screen.

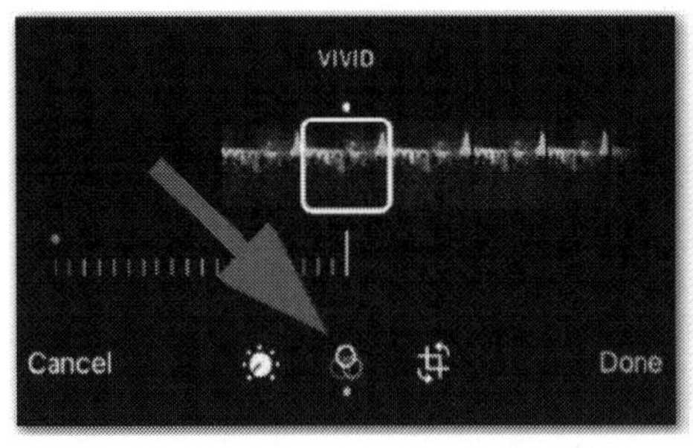

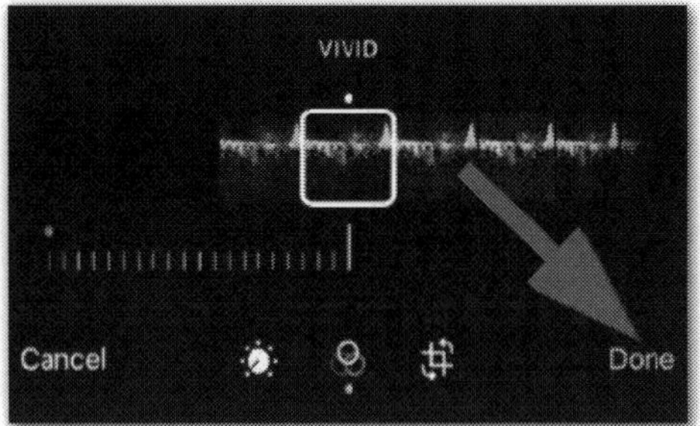

Scan a QR code

Quick Response (QR) codes are machine-readable bar codes that store information, such as website URLs, email addresses, etc. iPhone's camera can scan QR codes to provide quick access to online services, such as paying bills, opening a webpage, or sending information. Use iPhone's Camera app to scan a QR code.

1. Go to **Camera**.

2. Switch to the back camera and place it on the QR code you wish to scan. Make sure the code fits your Camera screen. When the camera recognizes the code, it will show a notification on the screen. Tap the **notification** to redirect you to the website or app.

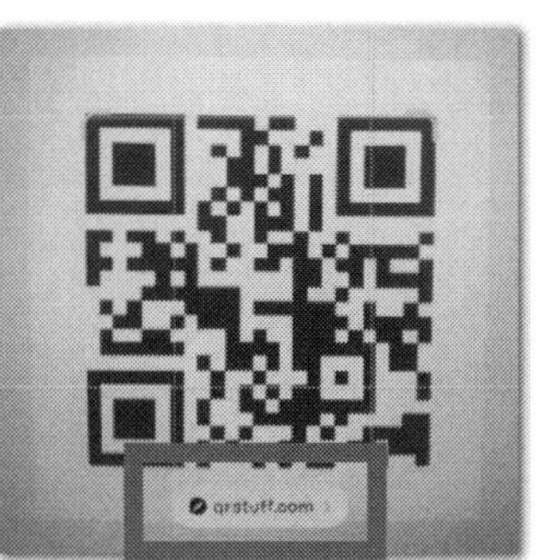

CHAPTER 07

WEB AND IPHONE APPS

pps are the backbone of the iPhone and, by extension, of any smartphone. With the help of these rograms, users may quickly and simply do their tasks. Add, edit, and navigate features quickly, and move eamlessly between applications with a wide range of personalization options—all on your iPhone. In this hapter, you'll learn several tips and tricks to get the most out of all the important apps on your phone.

3rowser

ou can use the Safari app to access the web, view websites, preview links to websites, translate web ages, and restore the Safari app to your Home Screen if you accidentally delete it. By using the same pple ID to sign into iCloud on different devices, you may synchronize your browser tabs, bookmarks, story, and Reading List.

avigate websites with Safari

earch for a specific website

Go to **Safari**.

2. Tap the **Address** bar at the bottom of your screen, then input the website or URL you want to visit.

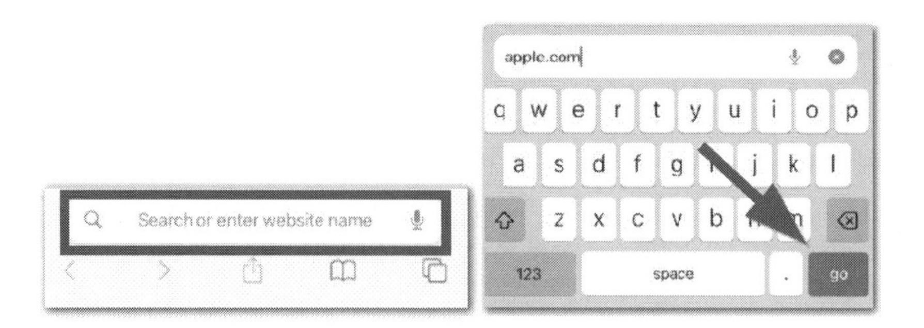

Add a website to Reading list, Bookmark, or Favorites

1. Tap the **share icon** at the bottom of your screen, scroll down and select from any options displayed

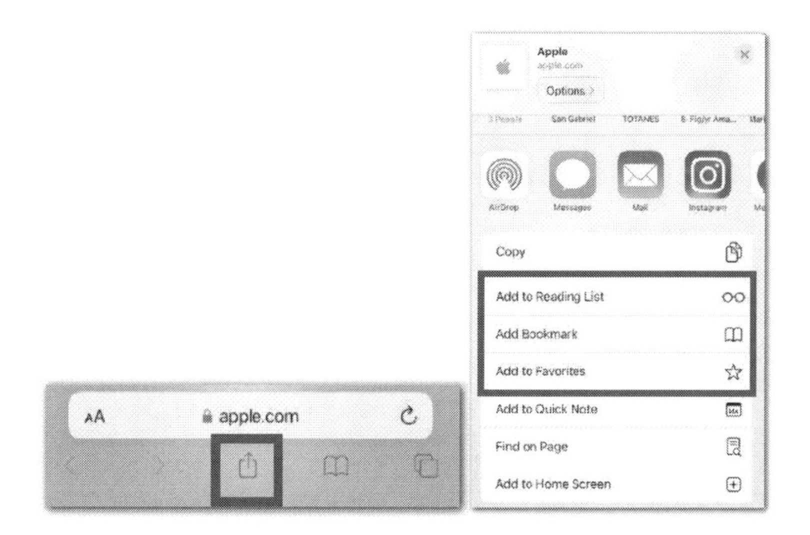

View saved websites as Reading list, Bookmark, and History

1. Tap the **book icon** at the bottom of your screen and select from the displayed icons what you wis to view.

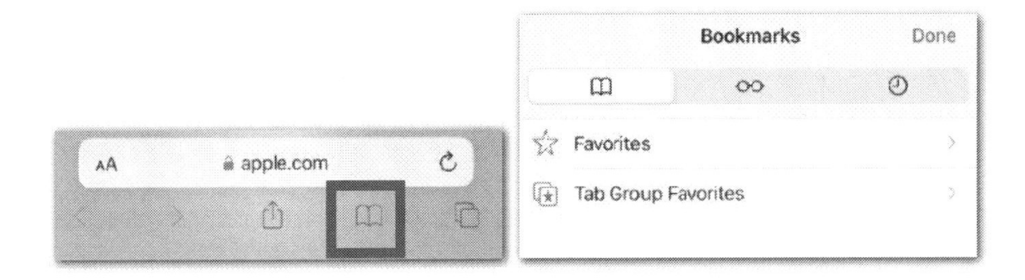

2. Tap **Done** at the top-right corner of your screen to go back to the webpage.

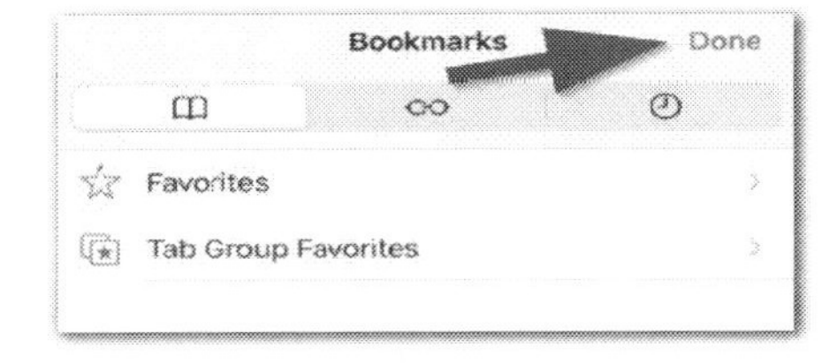

See more of the page

1. Switch on the iPhone landscape orientation by accessing the control center and tapping the **Lock** icon to unlock.

Refresh the page

2. Swipe down from the top of the page and the loading icon will show. Wait for it to disappear.

Share links

1. Tap the **Share** icon at the bottom of your screen, then share it using the desired option.

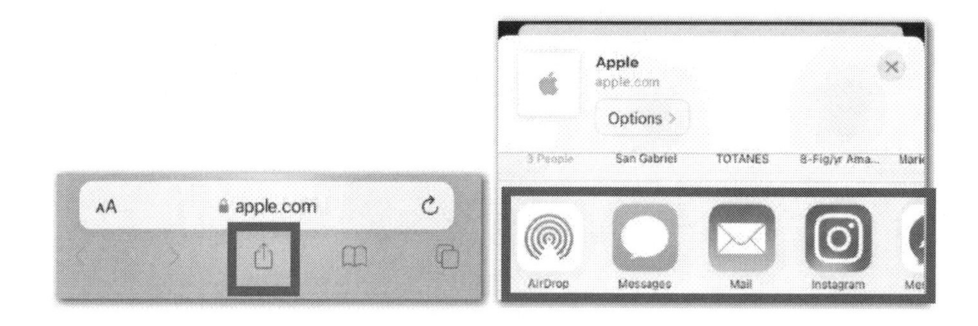

Preview websites and links

1. You can preview a link sent to you on apps. Touch and hold the link received to preview the websit
   You may also select any desired option.

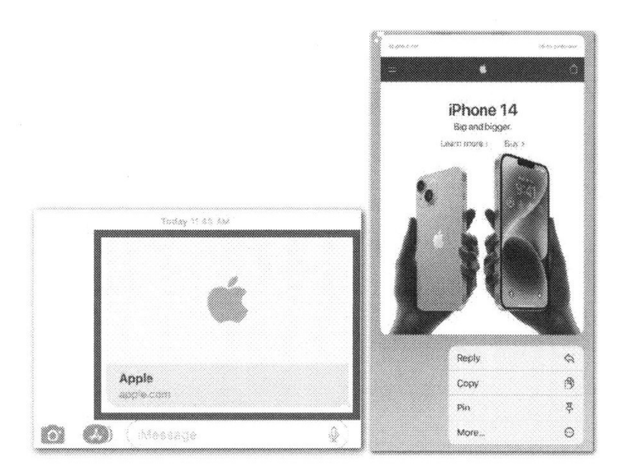

Restore Safari to the Home screen

Safari can be re-added to the Home Screen via the App Library if it has been removed.

2. Go to Home Screen, and swipe left until you see the App Library. Enter Safari in the search field.

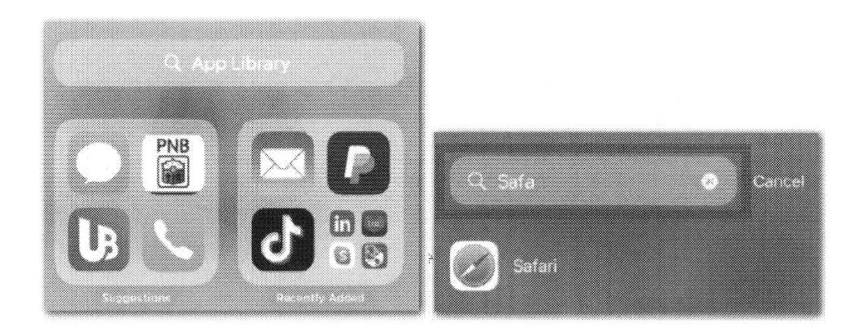

2. Touch and hold the **Safari icon**, then tap **Add to Home Screen**.

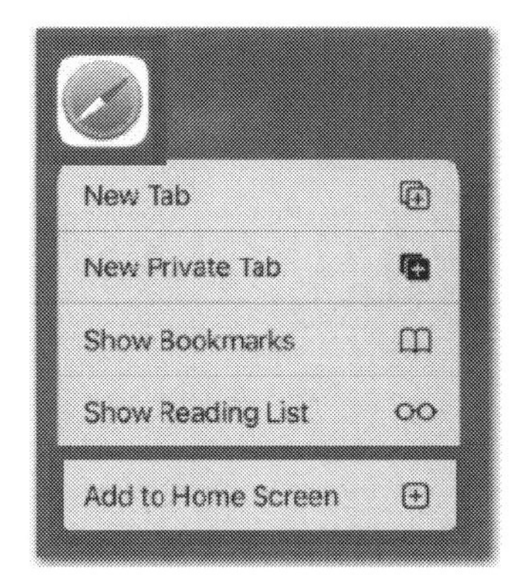

Tip: You can also download the Google Apps to your iPhone.

1. Go to **App Store**.

2. Tap **Search** at the bottom of your screen and type the Google application you want to download.

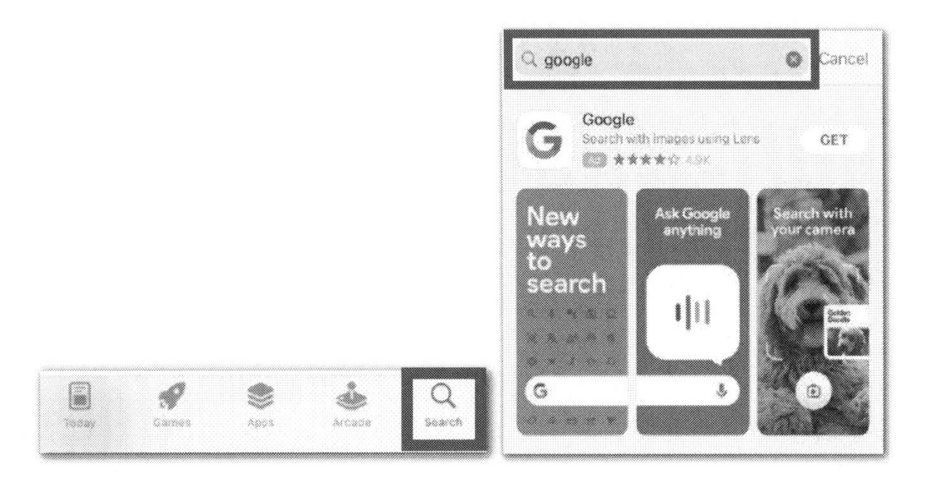

3. Tap **Get**, then tap **Install**. If prompted, enter your Apple ID password, or proceed with verification such as double-clicking the side button or Touch ID (for iPhone SE).

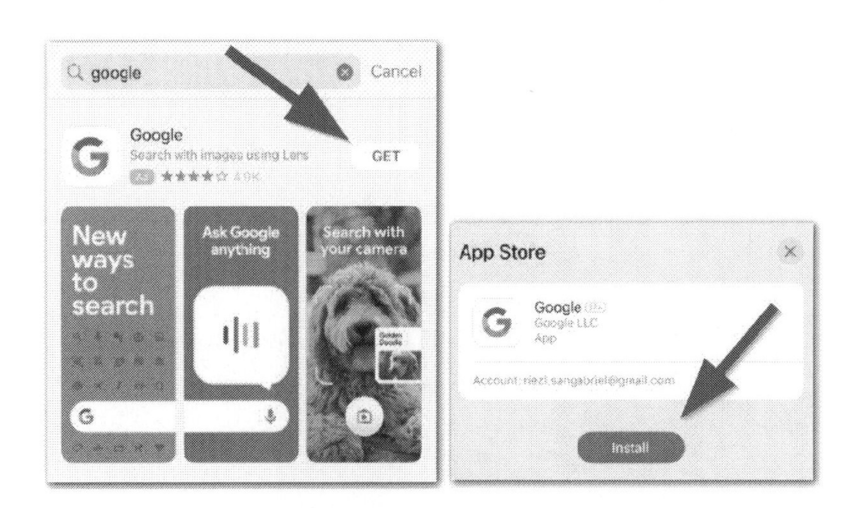

You can download more apps using App Store. More information about app download and installation will be covered in the latter part of this chapter.

iPhone Apps

Mail

Use Mail to accomplish anything you would do with an email client, including reading and composing new messages, responding to existing ones, and organizing your inbox.

Add Email to Mail app

1. Go to **Settings**.

2. Scroll down or type **Mail** on the search bar. Tap **Mail**.

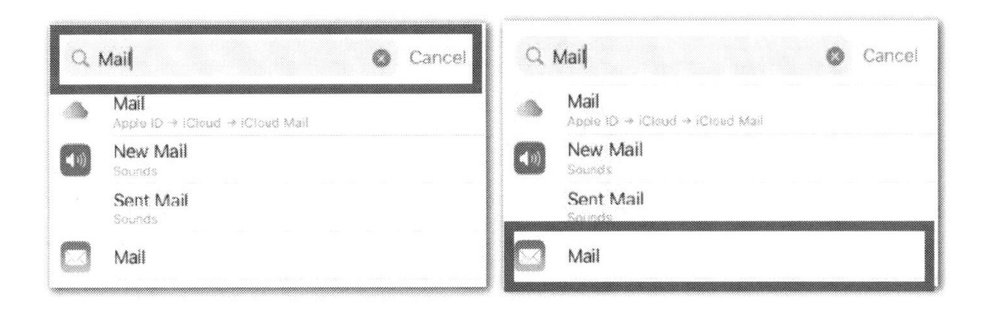

3. Tap **Accounts**, then Tap **Add Account**,

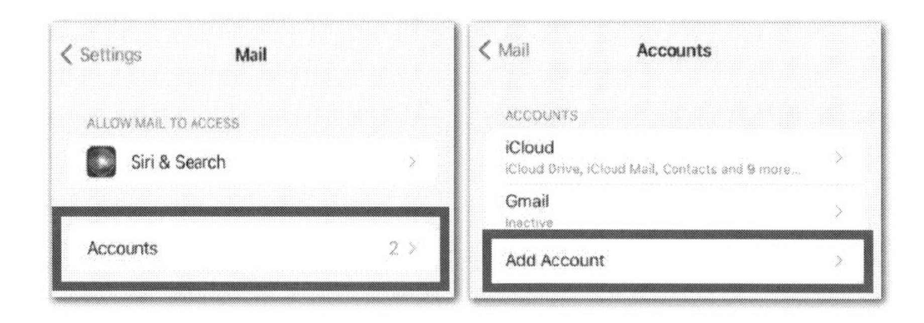

4. Select your email provider. Enter your email address and password, then tap **Next**.

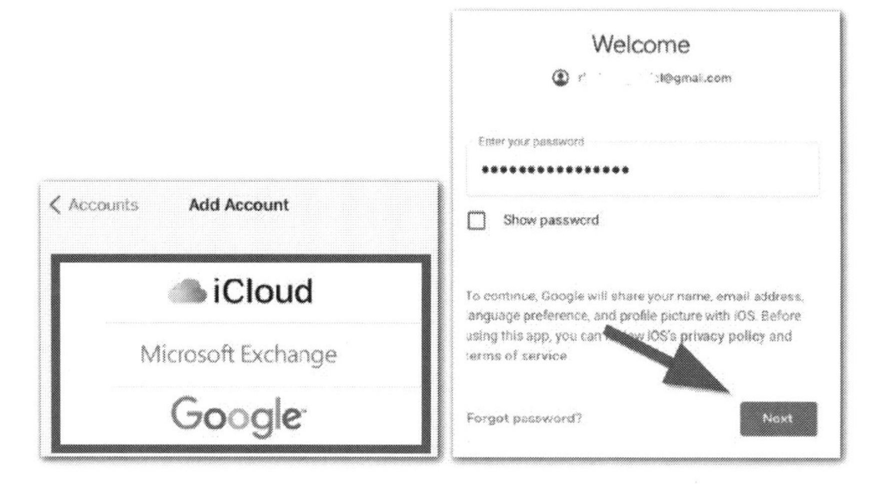

5. Tap **Allow** to allow iOS access to your account and wait for your account to be verified. Onc done, tap **Save**.

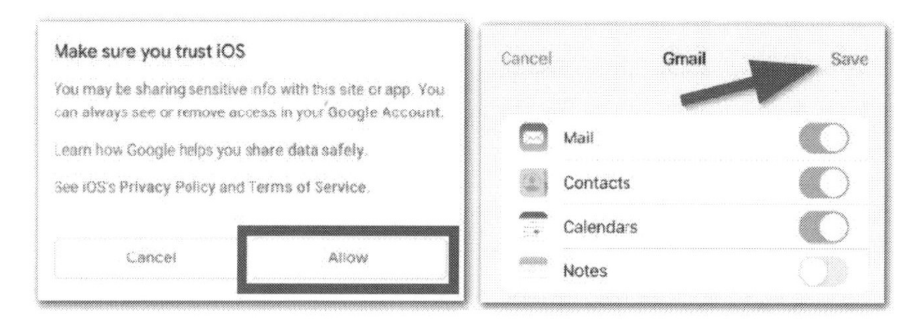

Maps

The Maps app has many uses beyond just locating a specific location or business. The Maps app allows you to pinpoint your current location on a map and then zoom in or out to view as much or as little of the surrounding area as necessary.

Find a location

1. Go to **Maps**.

2. If Maps alerts you to use your location, tap **Allow Once** or **Allow While Using the App**. Selecting Don't Allow won't let you access the app at best.

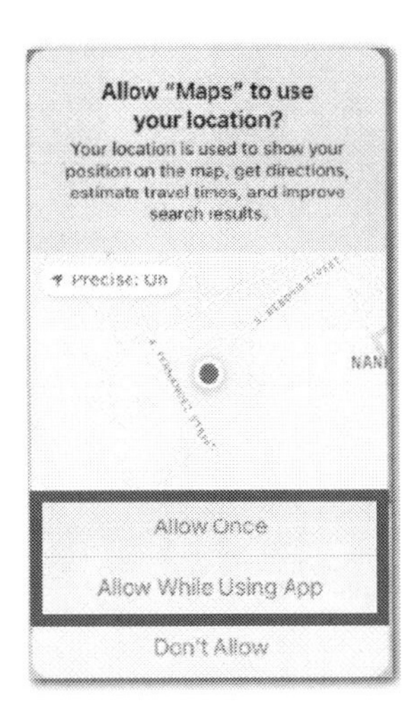

3. Tap the search field and type the desired location you want to look up. You can tap a suggestion for a faster search.

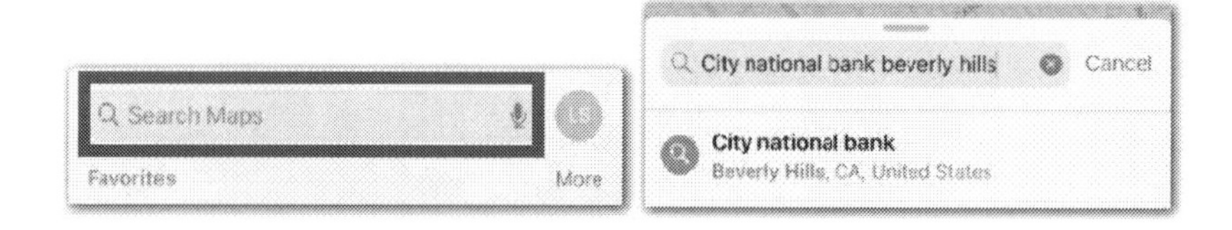

4. Tap **Search**.

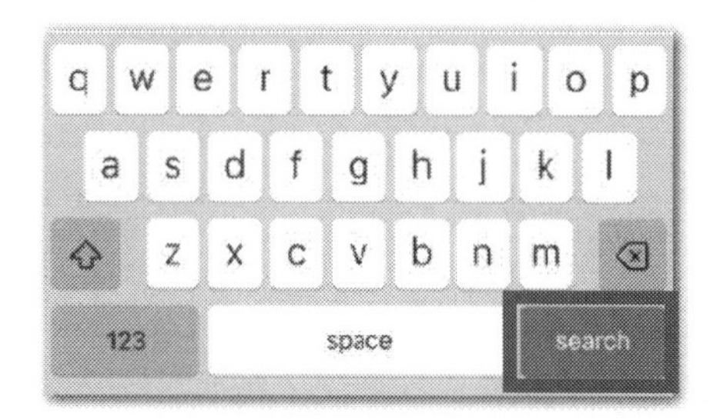

Let Maps utilize your precise location

Your iPhone needs to be connected with the internet and have Precise Location turned on to pinpoint you location and deliver precise directions.

A. If Maps alerts you that Location Services is disabled, enable it by:

1. Go to **Settings**.

2. Scroll down or type **Privacy & Security** on the search bar. Tap **Privacy & Security**.

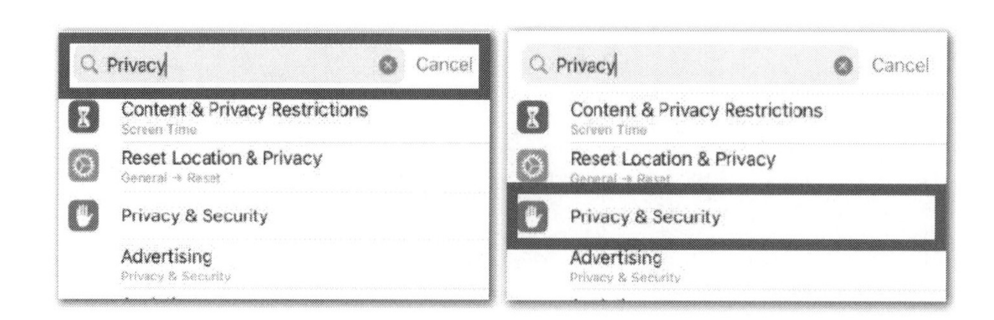

3. Switch on **Locations Services**.

B. If the Precise Location warning appears in Maps, enable it by:

1. Go to **Settings**

2. Scroll down or type **Maps** on the search bar. Tap **Maps**.

3. Tap **Location Services**, then switch on **Precise Location**.

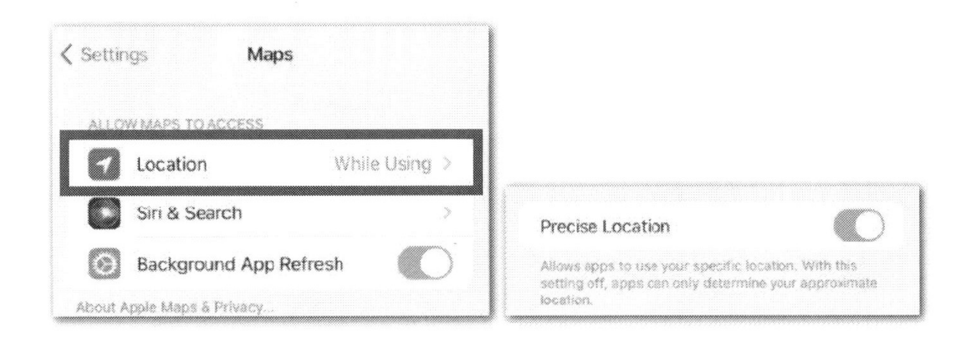

Get directions

You may get travel directions via the Maps app in different ways.

1. To search for a location, type the destination in the Search field, then tap a result.

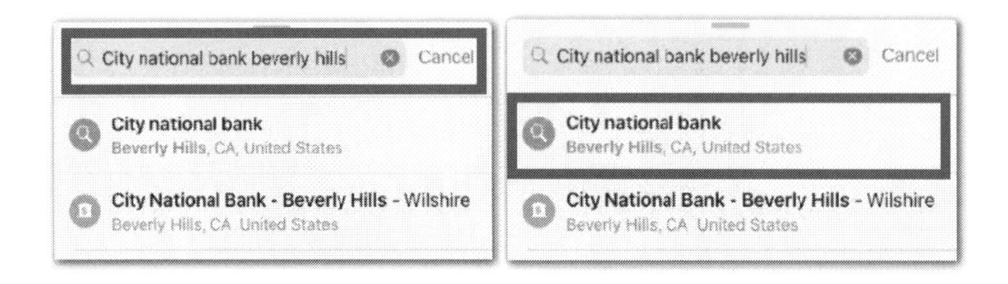

2. Tap the **directions button** on the place card. The result will show the direction from your locatior to the place you searched, but you can also choose a different starting point, a different mode o travel, and other options.

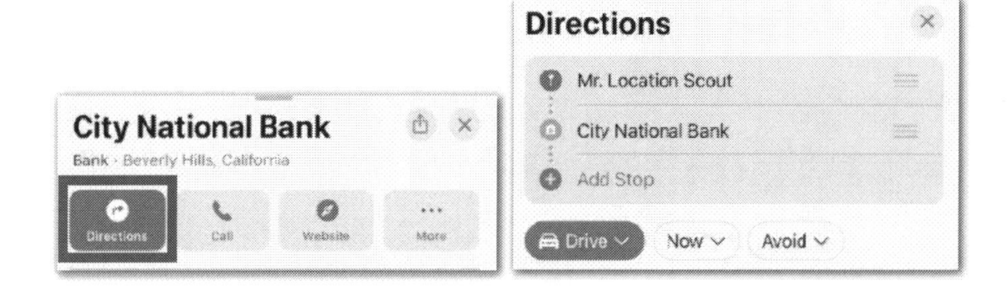

3. Scroll down to view and select alternative route options in Maps. Tap **Go** for the desired route. Ta the **location icon** to get directions as you move.

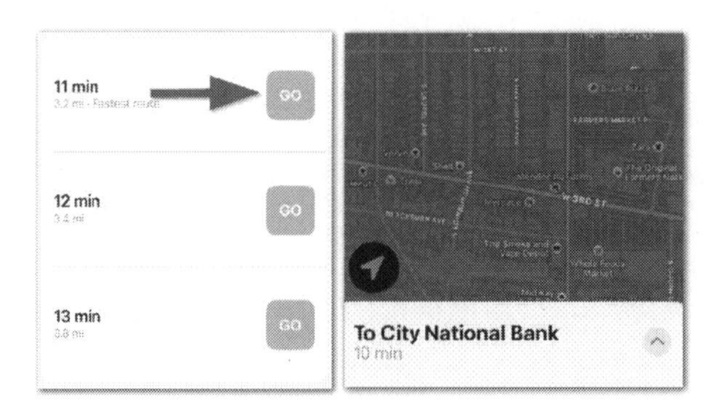

4. Tap **Details** to view and get an overview of the directions. To finish, tap **End Route**.

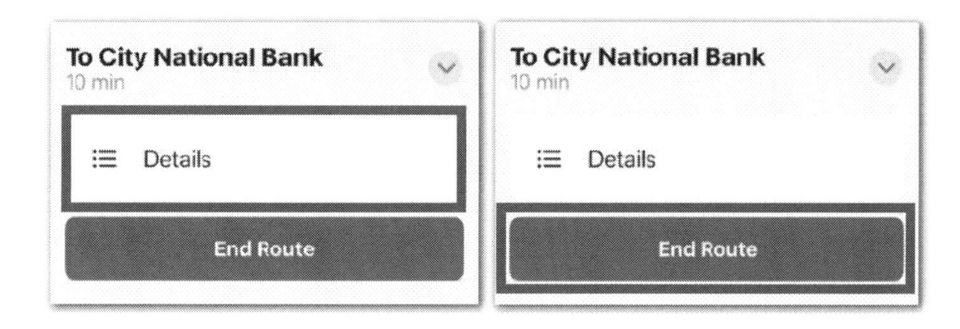

Mark locations with Pins

You can mark places in the Maps app with pins to help you find those places later. To mark your location:

1. Go to **Maps**.

2. Touch and hold the **Maps icon** on the Home Screen, then tap **Mark My Location**.

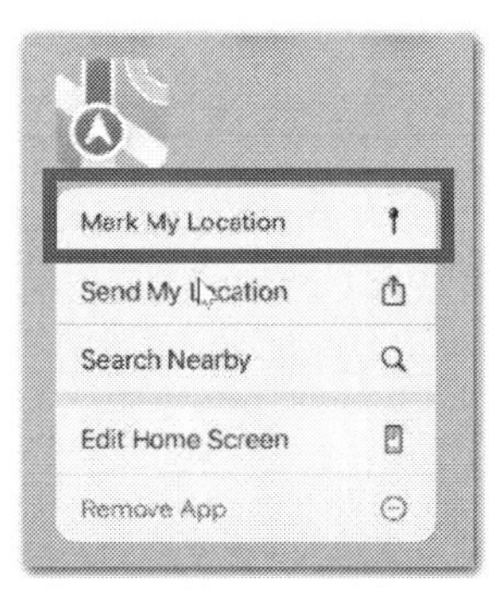

Delete a pinned location

Go to **Maps**, then tap the **marker**. In the place card, scroll down, then tap Remove.

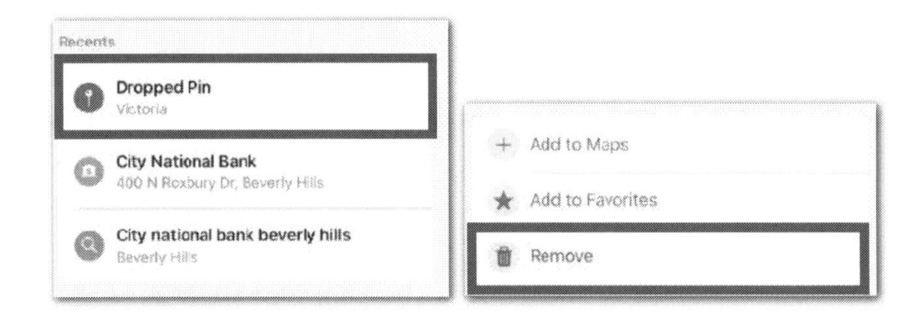

2. Alternatively, you may touch and hold the marker, then tap Remove Pin.

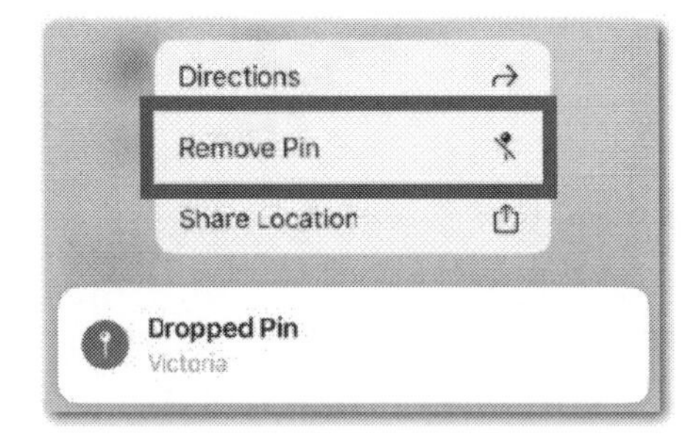

Clock

In the Clock app, you can choose to set alarms at any time of the day and put them on repeat anytime an(any day.

Set a regular alarm

You may set regular alarms at any time, including one for the time you wish to get up. A regular alarm i unconnected to any sleep schedule.

1. Go to **Clock**.

2. Tap **Alarm** at the bottom of your screen, then tap the **Add** icon.

3. Set the time, then select any of the following options:

- To repeat – select the days of the week
- To add a label – give an alert label, such as "Drink water."
- To add sound – select a vibration, music, or ringtone
- To snooze – have more time before the alarm

4. Then, tap **Save** at the top-right corner of your screen.

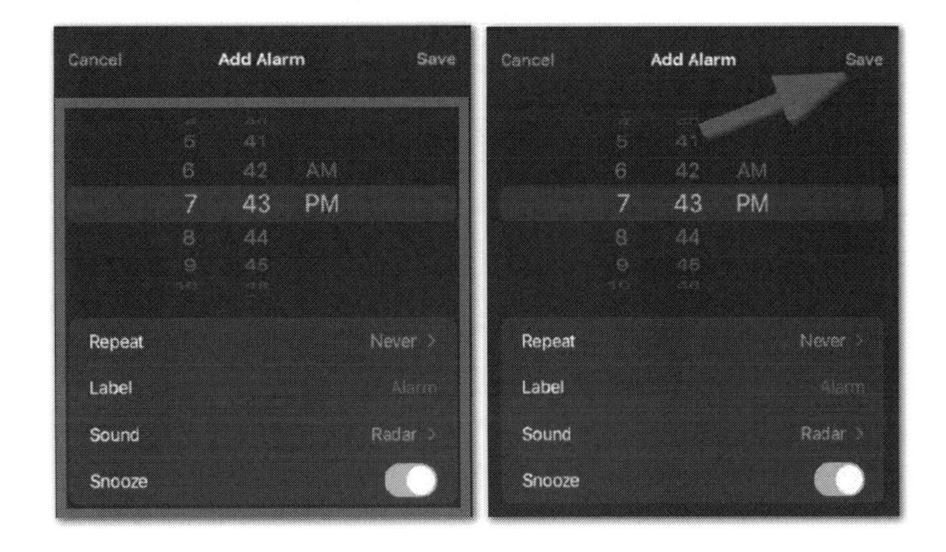

dit and delete alarm

To change alarm settings, tap **Edit** at the top-left of the alarm screen, then select the alarm time you want to edit.

2. Edit the alarm details, then tap **Save** at the top-right corner of your screen.

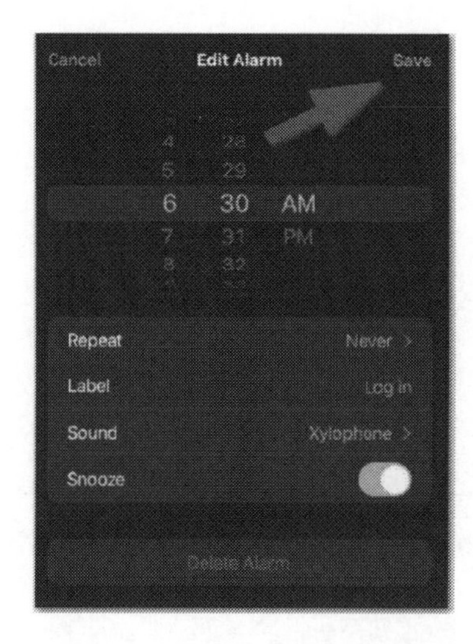

3. To remove an alarm, tap the **delete icon** near the alarm you want to remove, then tap **Done** at th top-left corner of your screen. Alternatively, you can tap the alarm, then tap **Delete Alarm**.

e the time in different cities and different time zones.

see the local time in different time zones across the world:

Go to **Clock**.

Tap **World Clock**. Tap the **Add** icon at the top-right corner of your screen to add extra clocks.

Select a city.

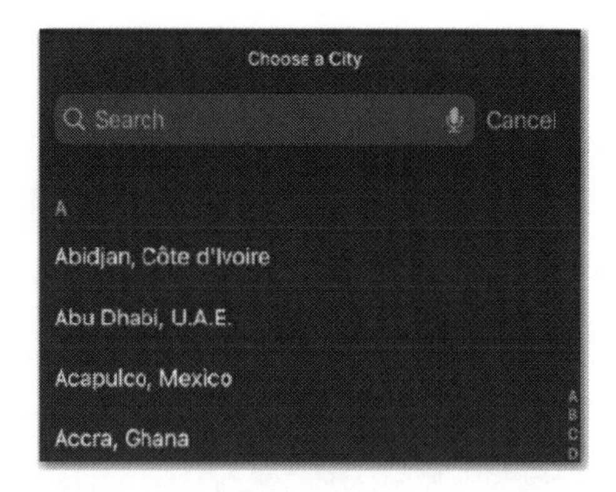

4. Tap the **Edit** button at the top-left corner of your screen to reorganize or delete clocks.

Manage your list of cities

1. On the World Clock screen, tap **Edit**.

➲ To add a city – tap the Add button, then pick a city

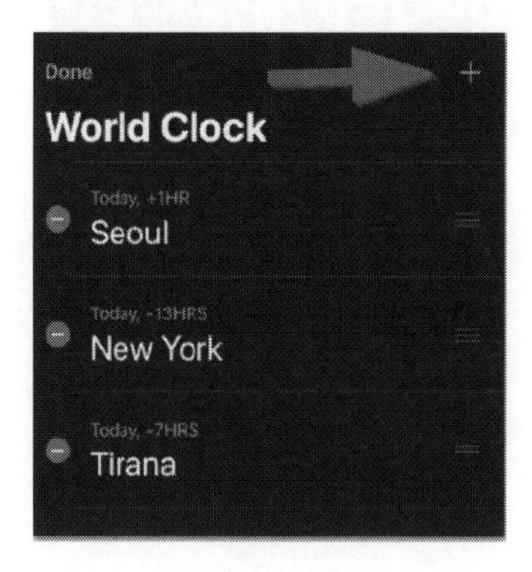

➲ To delete a city – tap the **Delete icon**, then tap **Delete**.

➲ To reorder the cities – drag the Reorder button up or down.

Stopwatch and Timer

rack time with the Stopwatch or Timer.

. Go to **Clock**.

2. Tap **Stopwatch** or **Timer**.

➲ To flip between the faces, swipe the stopwatch, then tap **Start**.

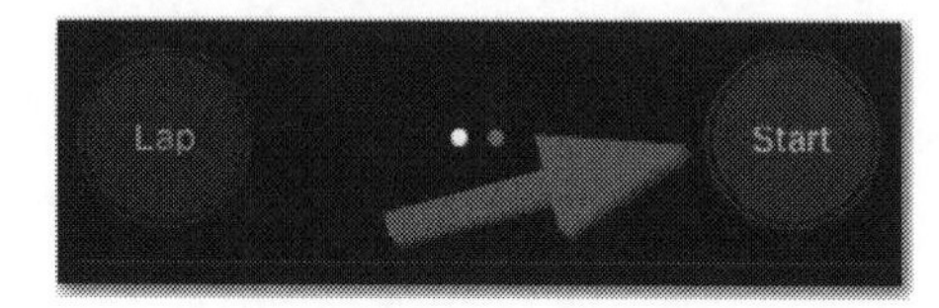

➲ To record a lap or split, tap **Lap**.

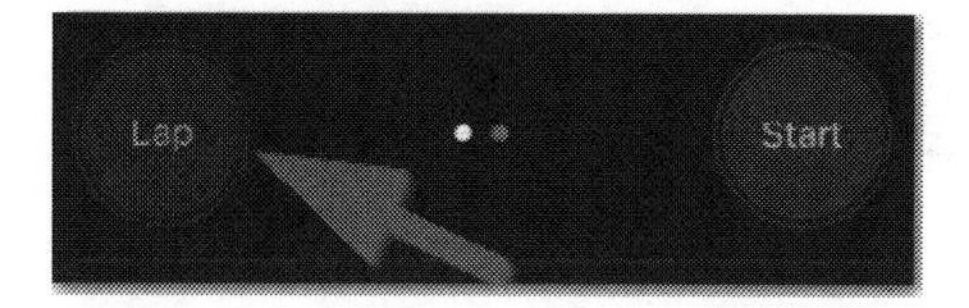

➲ Tap record the final time tap, **Stop**.

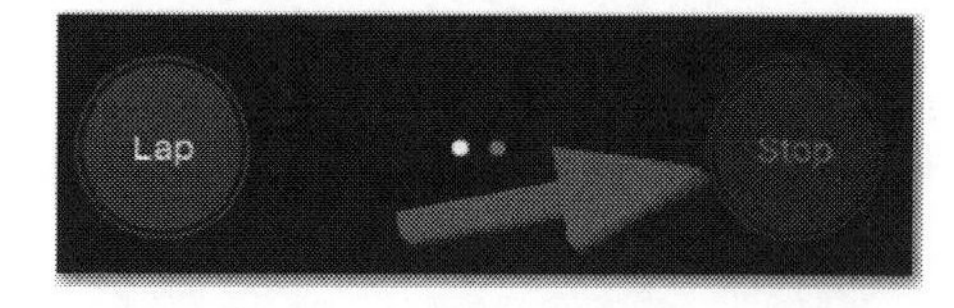

➲ Tap clear the stopwatch and tap **Reset**.

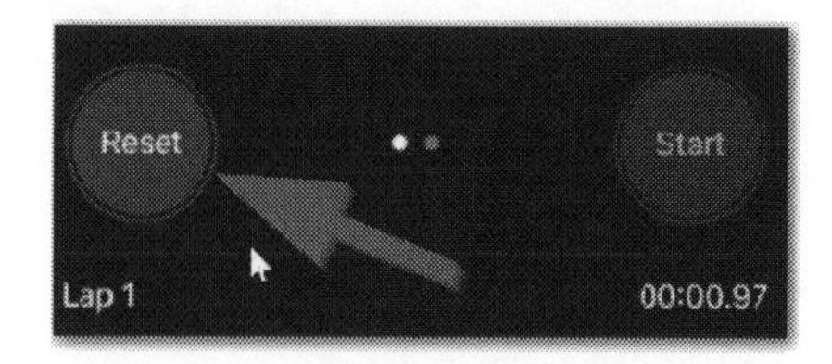

➲ To set the timer, tap **Timer**.

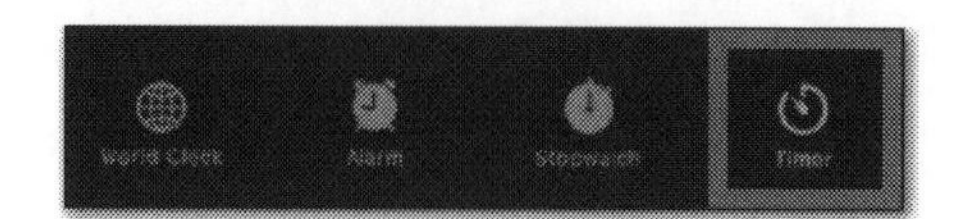

➲ To start the timer, tap **Start**.

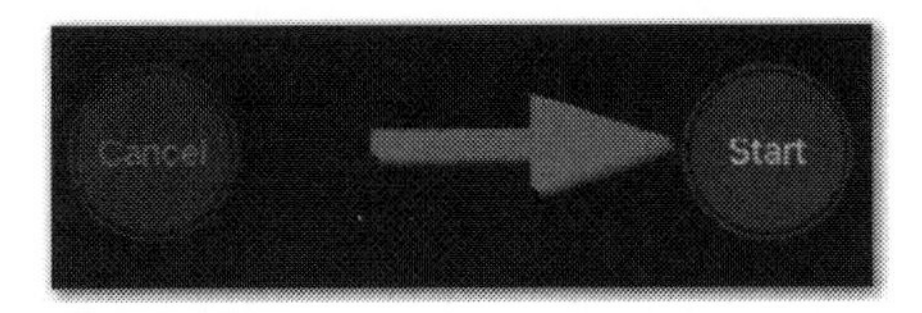

You may Pause, Resume or Cancel the timer.

Calendar

Calendar customization

Calendar's settings allow you to change the default start day of the week, show week numbers, switch to different calendar (such as Chinese or Hebrew), override the system's time zone, and much more.

1. Go to **Settings**.

2. Scroll down or type **Calendar** on the search bar. Tap **Calendar**.

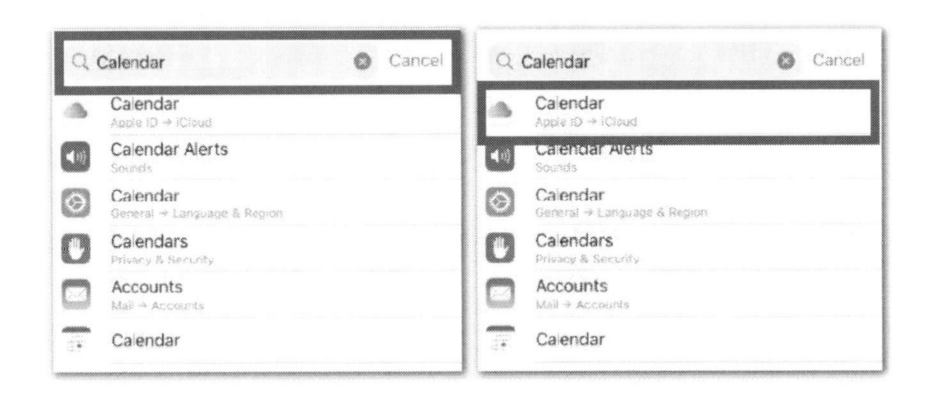

3. Navigate through the menus.

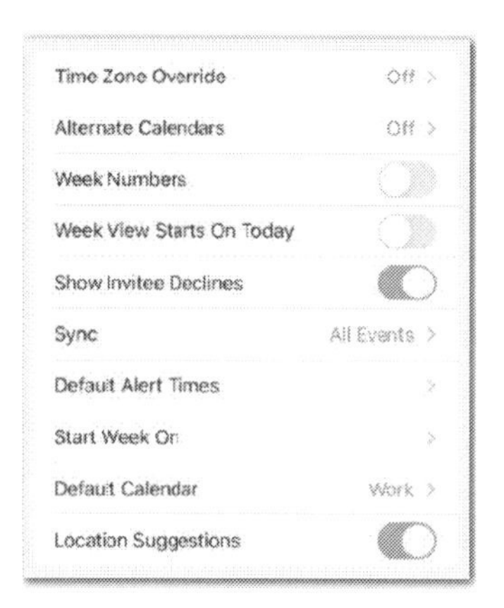

Create and edit an event

The Calendar app may help you organize and alter your day-to-day plans, as well as set up and attend important meetings.

Adding an event

1. Go to **Calendar**.

2. Tap the **add icon** at the top of the screen to add a new event. In the text field, type the event name

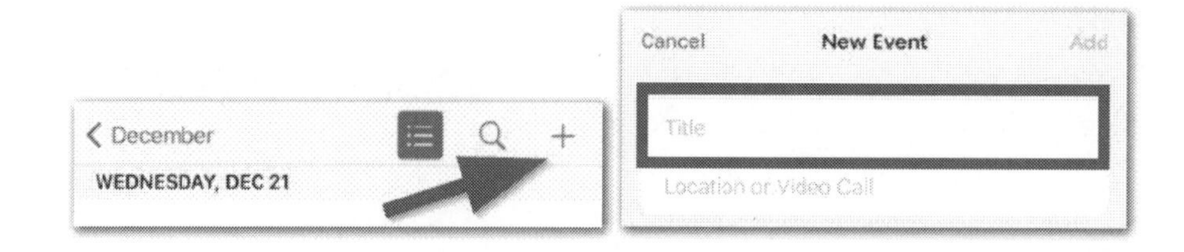

3. Tap the **Location or Video Call field** to enter a specific address or video call link to enter a video connection. If you have a FaceTime link that you sent or received, you may also paste that into the Location area.

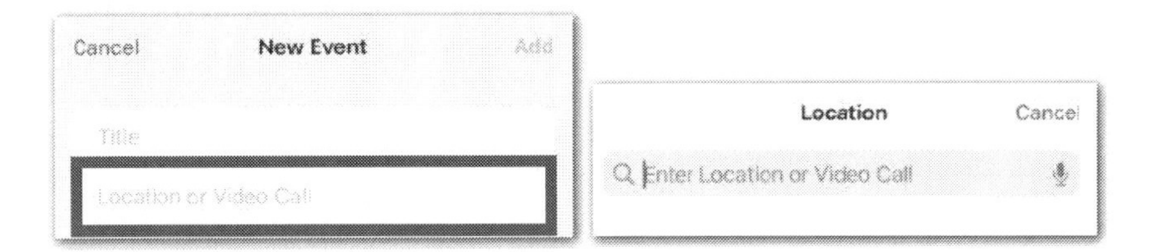

4. Dates and times, attendees, locations, documents, and more may all be added and selected in the options. Tap **Add** to finish.

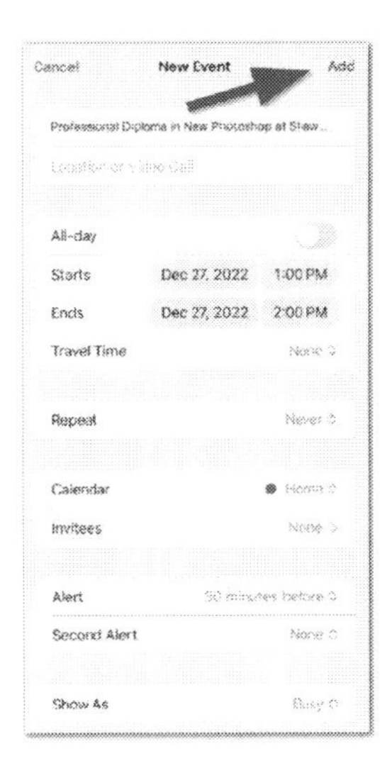

lter an event

event parameters, including time, are flexible and may be modified.

Tap a date to view events in the Day view, touch and hold the event, drag it, or reposition the grab points to move an event to a different time.

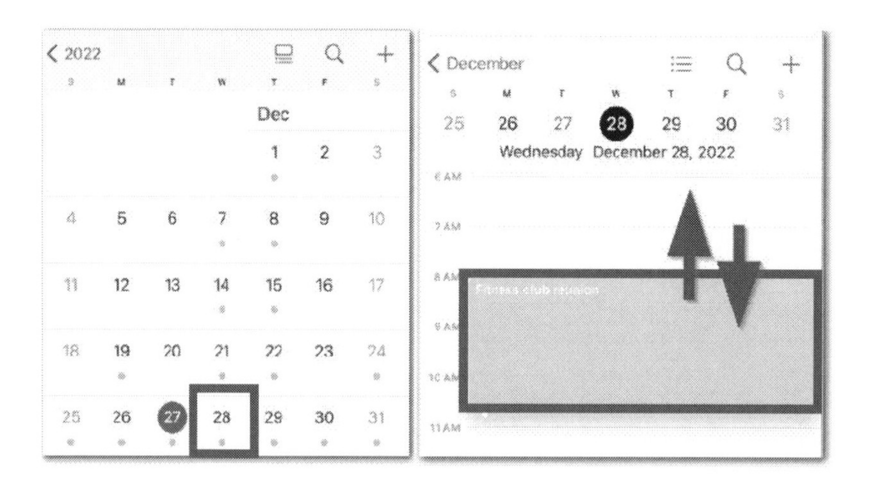

2. To modify an event's specifics, select an event, then tap **Edit** at the top-right. Tap a setting to modify it or tap in a field to enter new information. Tap **Done**.

Delete an event

1. To remove an event from the Day view, tap the event, then tap **Delete Event** at the bottom of your screen.

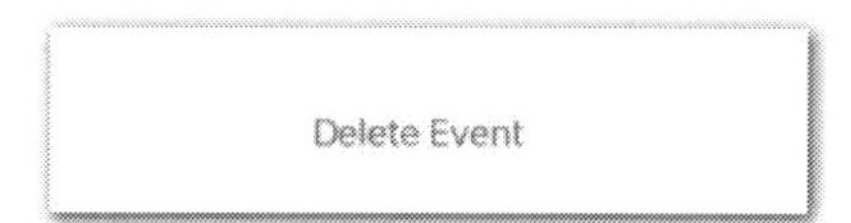

Reminders

Even though they serve similar purposes, calendars and reminders have a few key distinctions. Reminders are action-based events, whereas Calendar events are planned activities.

Create new Reminder

1. Go to **Reminder**.

2. Tap the **New Reminder icon** located in the bottom left corner of the screen.

3. In the text field, add reminder details, such as Title, Notes, and Details. Tap **Add**.

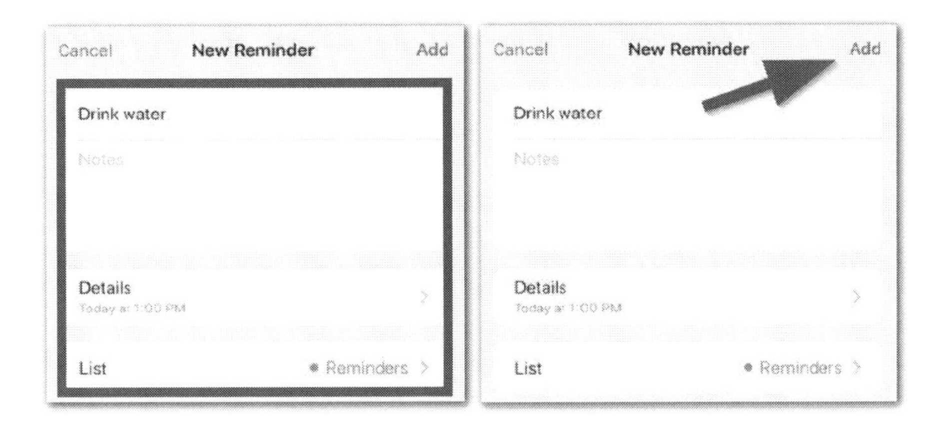

lanage a Reminder

'ith the Reminders app, it's easy to change and organize the things on a list.

dit multiple reminders

Tap **Reminders**, then tap the reminder you wish to edit.

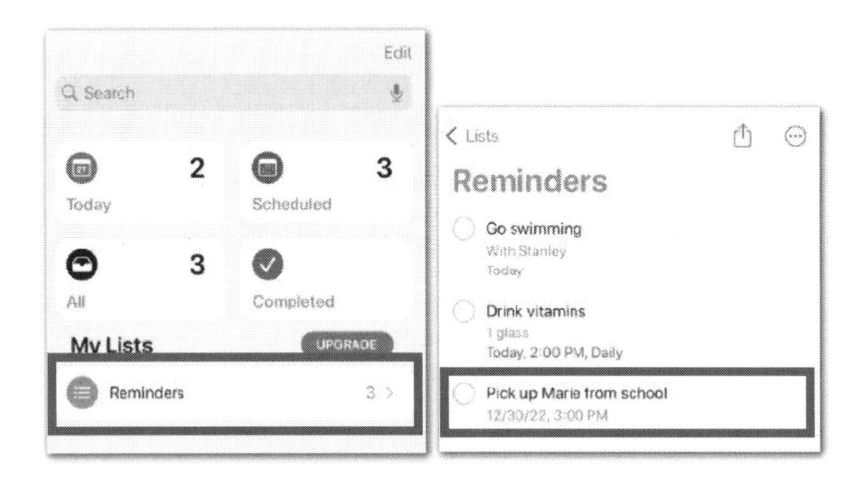

2. You can add a date and time or add a location to the selected items using the buttons at the bottom of your screen.

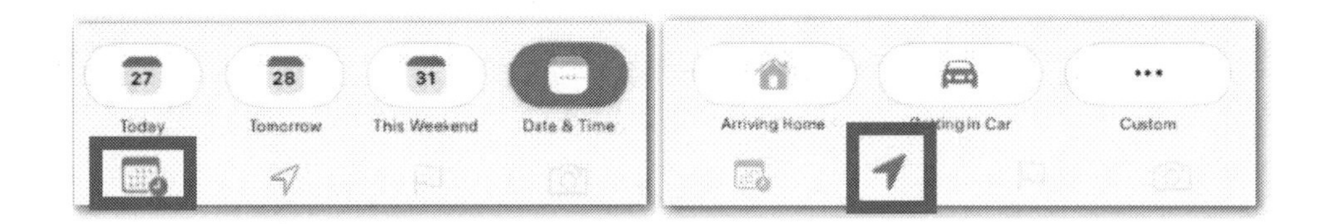

3. You can rearrange and sort reminders by looking at a list, tap and hold a **reminder** and move it around.

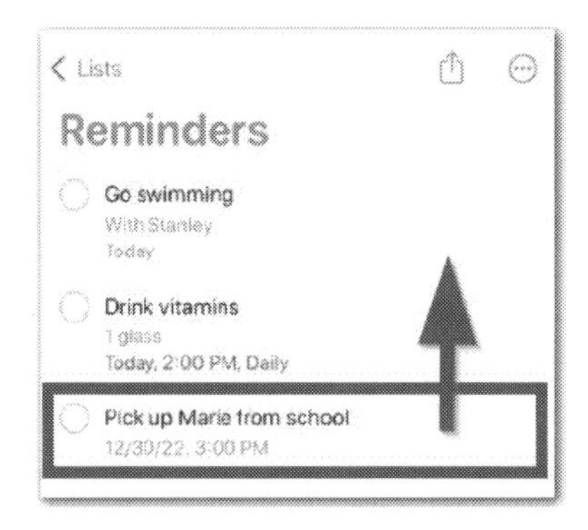

Check off items as done

1. If you tap the empty circle next to an item, it will be marked as done and disappear.

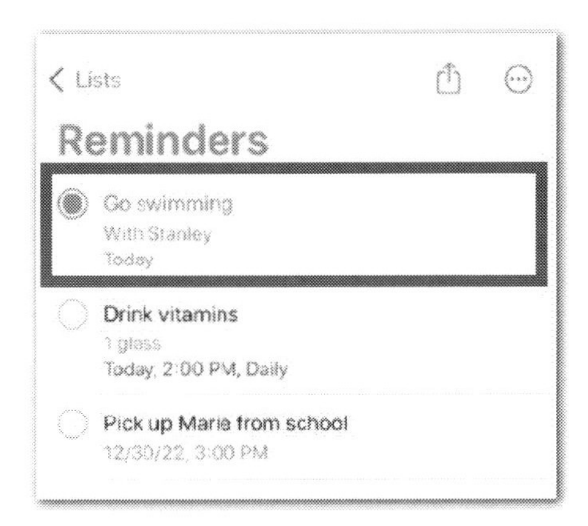

Delete a reminder

1. Swipe the item left, and then tap the **Delete** button.

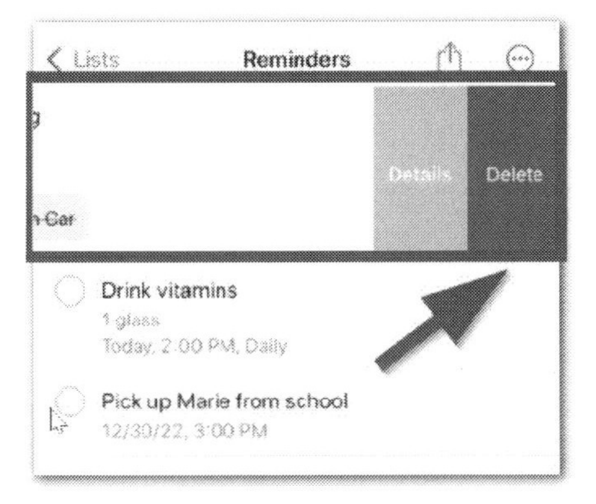

Notes

ou may jot down a few words, make a list, draw out some thoughts, and much more using Notes. iCloud ows you to sync your notes across all your gadgets.

reate a new note

Go to **Notes** app.

2. To write a new note, tap the **compose icon** at the bottom part of the screen. Finish writing notes by tapping the **Done** button.

Note: The note's title is the text of the first line.

To change the formatting style

1. Go to **Settings**.

2. Scroll down or type **Notes** on the search bar. Tap **Notes**.

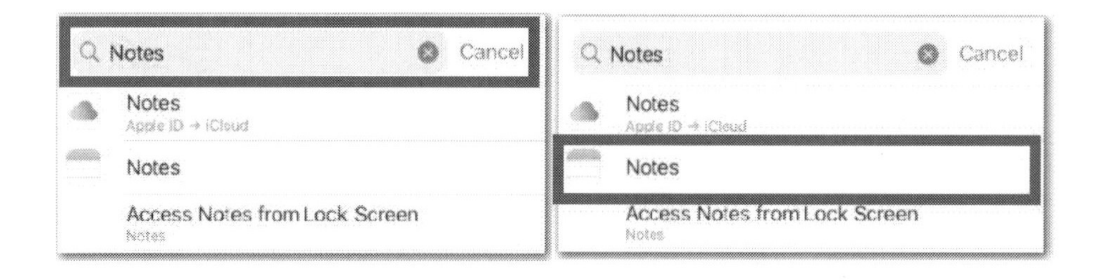

3. Tap **New Notes Start With** to modify the first line's formatting.

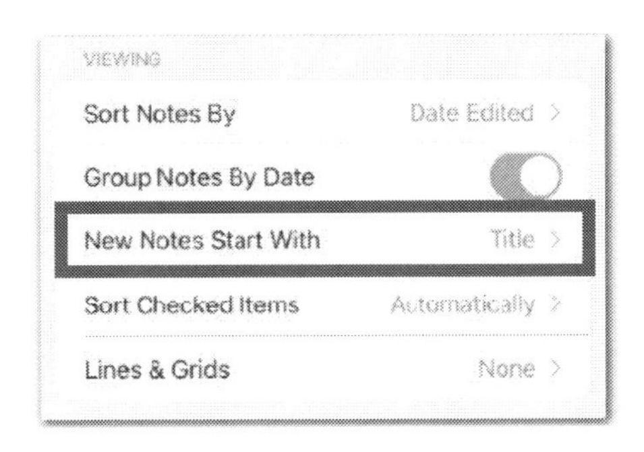

:dit a note

4. 1. Tap **Notes** from the Folders screen. Select the note you want to edit, then make changes.

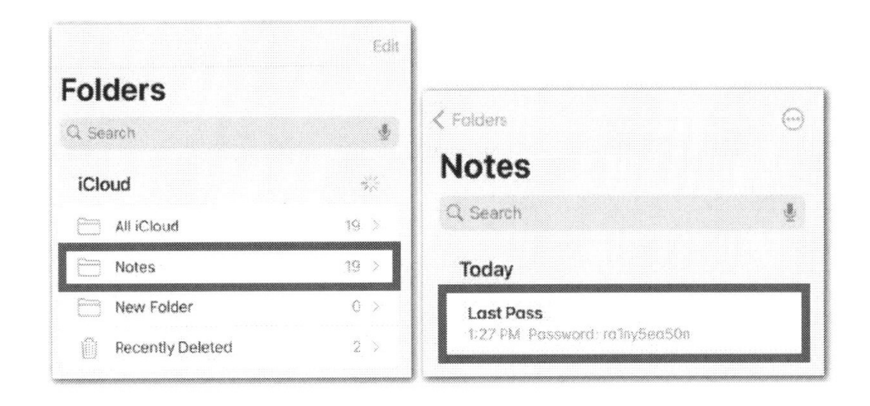

'elete a note

emove a note from your list

. Tap a **note**, then slide the finger left over the note. Select the **Trash** icon.

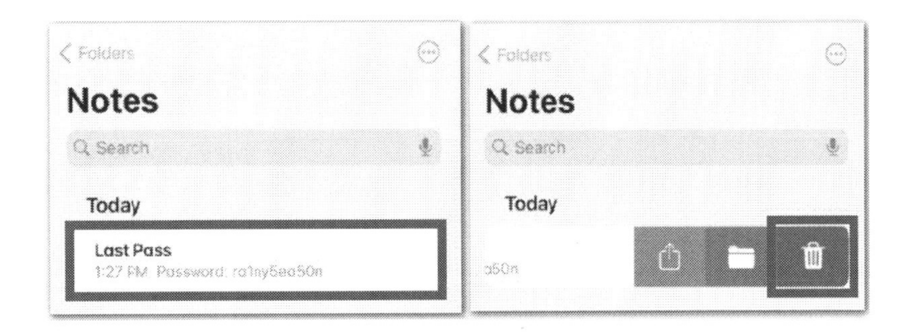

Retrieve a lost note

1. Select **Recently Deleted** from the Folders menu. Tap **Edit**.

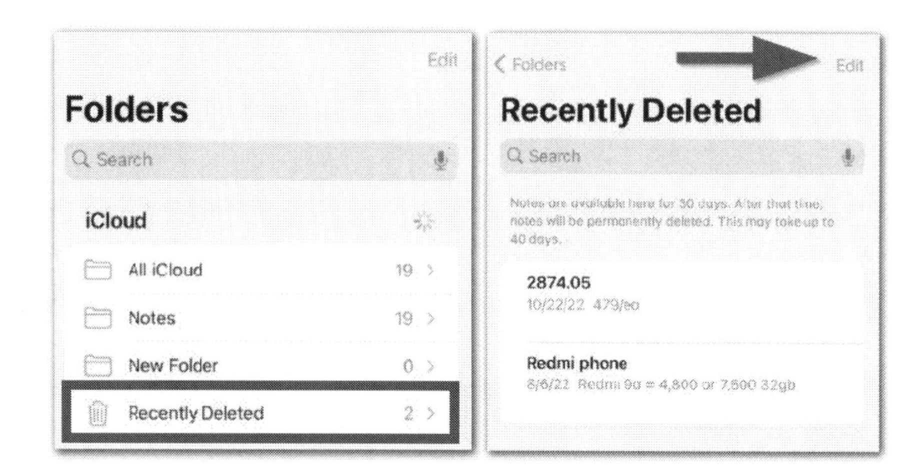

2. Select the note you want to retrieve. Tap **Move** at the bottom of your screen.

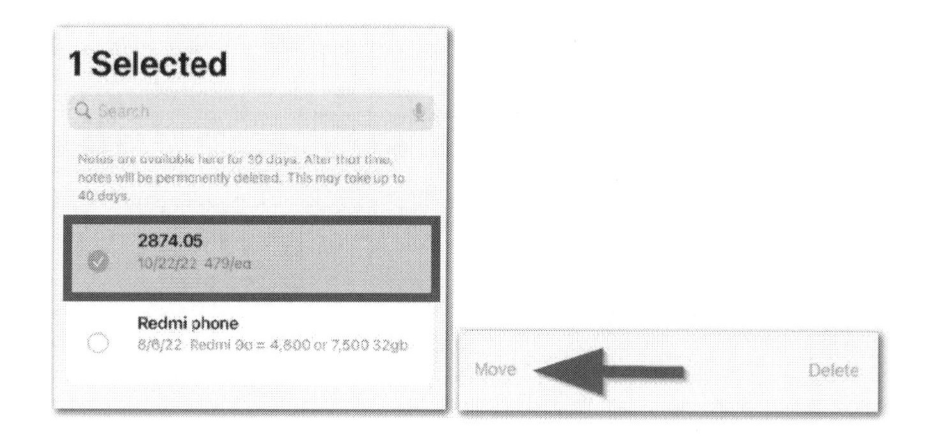

Listen to Music

You may use your Apple ID to login into Apple Music on any of your other compatible devices. All you mu do is follow the instructions for your specific device.

1. Go to **Apple Music**.

2. Tap **Listen Now**, then tap the **Photo** icon at the top-right corner, then follow the prompt on the screen to set up your account.

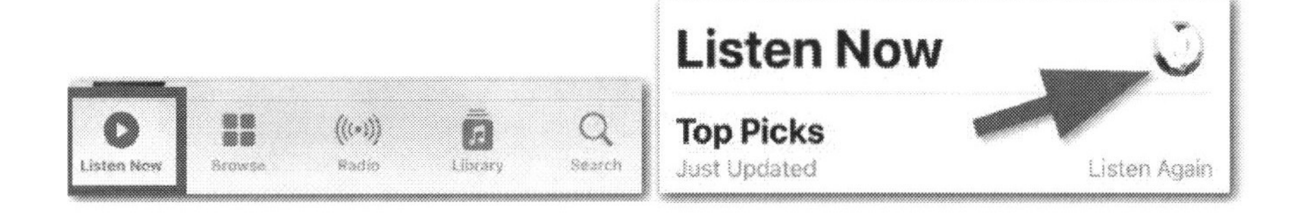

3. Choose to play songs by searching the title or artist in the **Search** field located in the bottom-right corner of your phone.

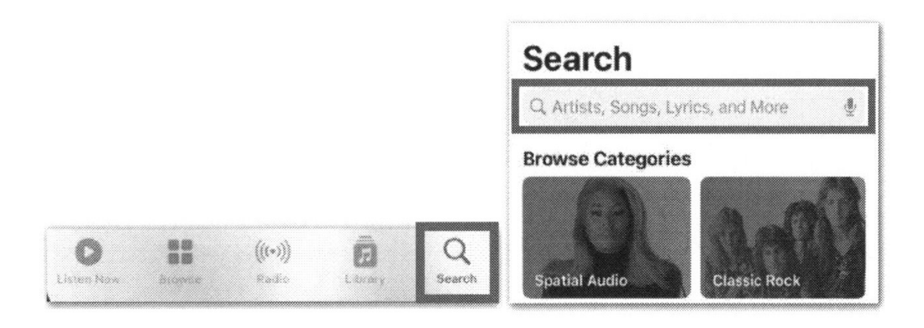

ou may also use other streaming services, such as Spotify and Deezer, to listen to songs and podcasts. hese applications are available in the App store. More information about app download and installation in ıe latter part of this chapter.

.ocating Apps and Switching Between Apps

nd out the quickest way to switch between apps. When you make the transition back, you may pick off ıst where you left off.

For iPhones 13 and 14

1. To access the App Switcher, swipe up from the bottom of the screen while you touch and hold until it appears in the center. Swipe left or right to select the desired app.

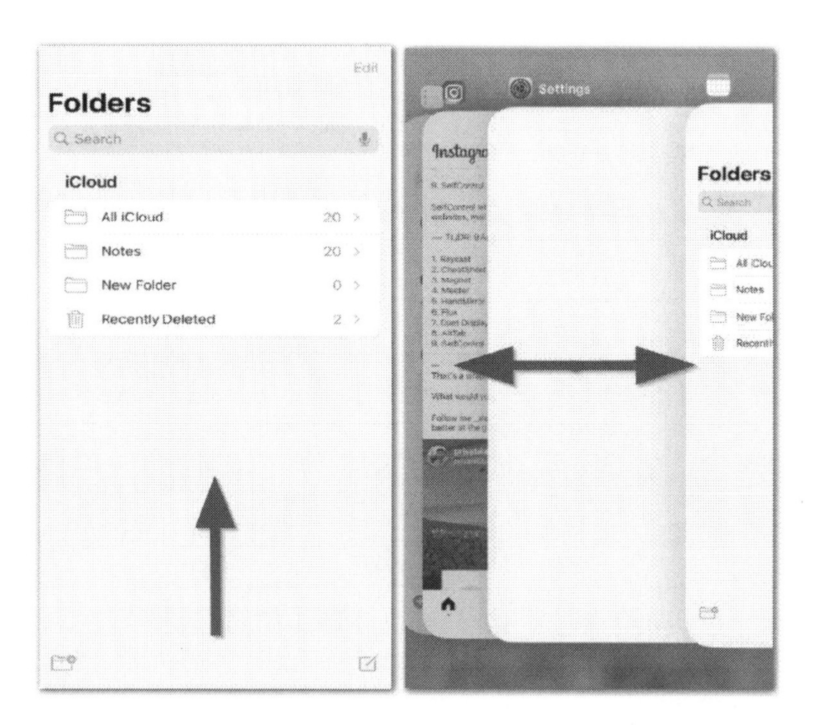

2. Tap the app you wish to view.

For iPhone SE

Switch an app with the Home button

1. To access the App switcher, just double-tap the **home button**. Swipe left or right to select a desire app, then tap the app you wish to view.

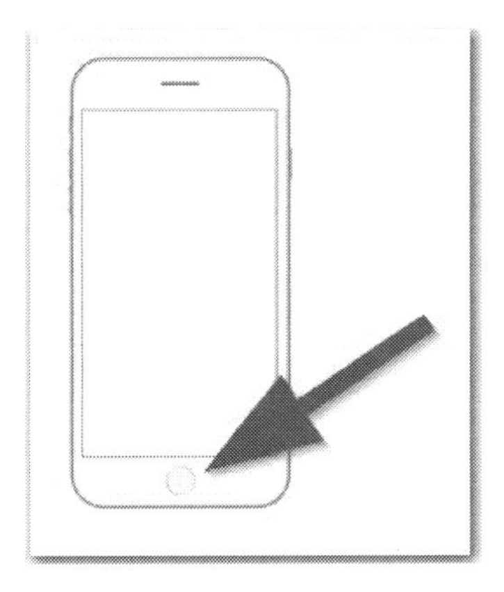

Delete an app

1. On your Home screen, touch and hold the app you want to delete. Tap **Remove App**.

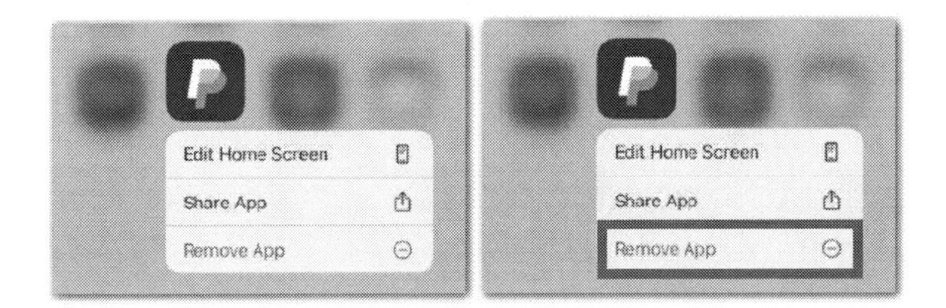

2. Select **Delete** App, then tap **Delete** to confirm.

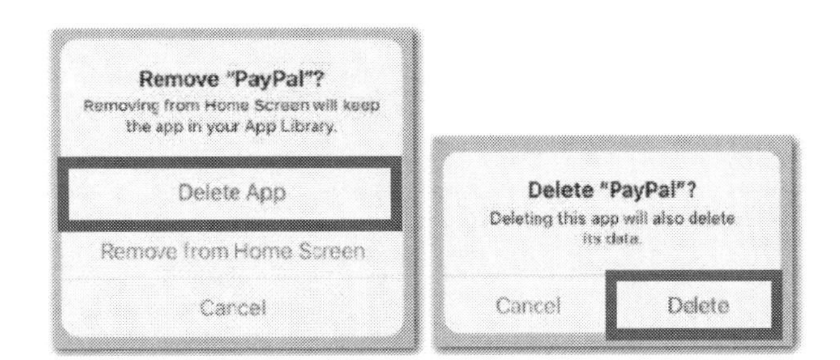

Find and launch an App

1. To access the App Library, launch the Home Screen and swipe left to get through all the Home Screen pages. Locate a certain app by typing its name into the App Library search bar. You may also use the scroll bar to move through the alphabetical list.

2. Tap on an icon to launch the associated app.

Tip: When browsing categories, look for groups of app icons; if there are just a few, touch on them to reveal all the applications in that group.

The App Store

Download applications from the App Store and install them on your iPhone.

1. Go to the **App store**.

2. Tap **Search** at the bottom of your screen and type the app you want to download. Select the app.

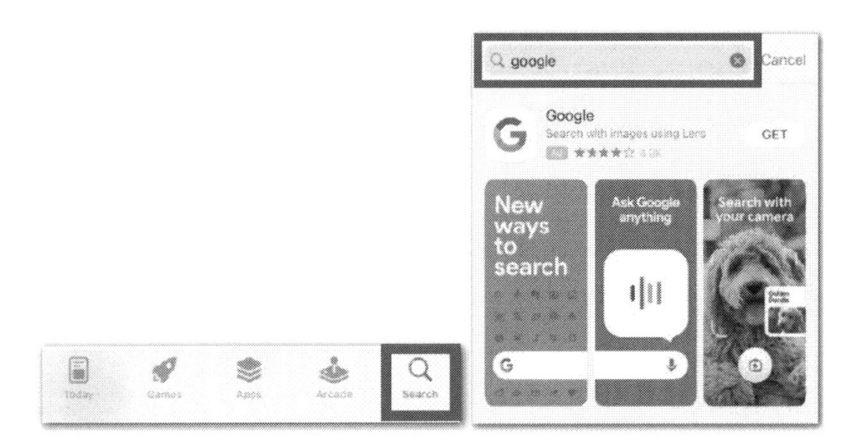

3. Tap **Get**, then tap **Install**. If prompted, enter your Apple ID password, or proceed with verification, such as double-clicking the side button or Touch ID (for iPhone SE).

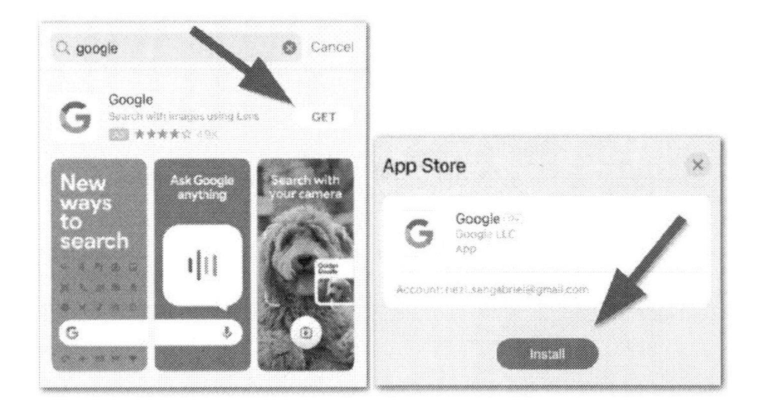

4. Successfully downloaded app will appear on your home screen.

CHAPTER 08

PRIVACY, SECURITY, AND OPTIMIZATION

Having a smartphone is fun and serves its purpose – making life easy. However, as much as you wan to maximize its efficiency, consider protecting it from harmful threats. Your iPhone has built-in privac and security features that you should be aware of. iPhone creates a safe phone experience for you an provides other useful functionalities that everyone, especially seniors, can take advantage of. Feature that help you in case of a device loss or emergency are already installed on your phone, so make sure t add them to your iPhone setup.

Built-in Security and Privacy Measures

iPhone's built-in security and privacy features are designed to enhance the safety of information stored o your phone within its environment. Notice how your device prompts you to set up security details; this is a example of the internal safety measures at work. There are more protective features that you can optimiz on iPhone, and for them to work as well as possible, follow these safety practices:

1. **Make sure to set a strong passcode.** Consider using a passcode that is not related to your person information and will be difficult for others to discover. Add a combination of characters like numbe and symbols to strengthen and remove weak passwords. Enable Face ID or Touch ID to add a added layer of protection from unauthorized individuals who may access your phone.

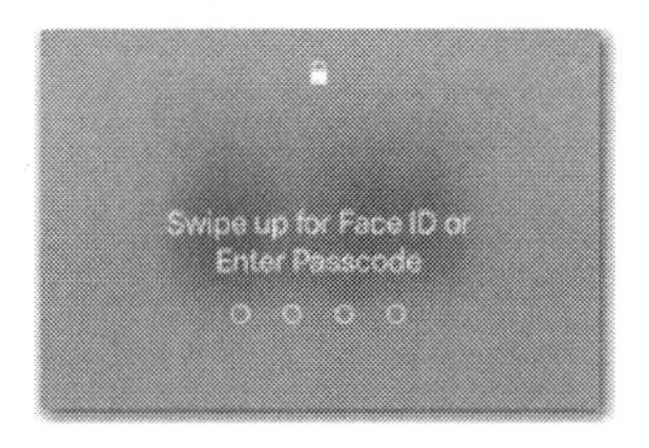

2. **Use iPhone security features when signing into apps and website accounts.** Face ID or Touch ID can be used as a passkey to sign into your accounts instead of entering your username and password. You can also create new accounts for apps and websites using your Apple ID, so you don't have to provide a unique username and password. Signing in via iPhone security features is secure and easy and limits the security information you need to remember and keep for different apps.

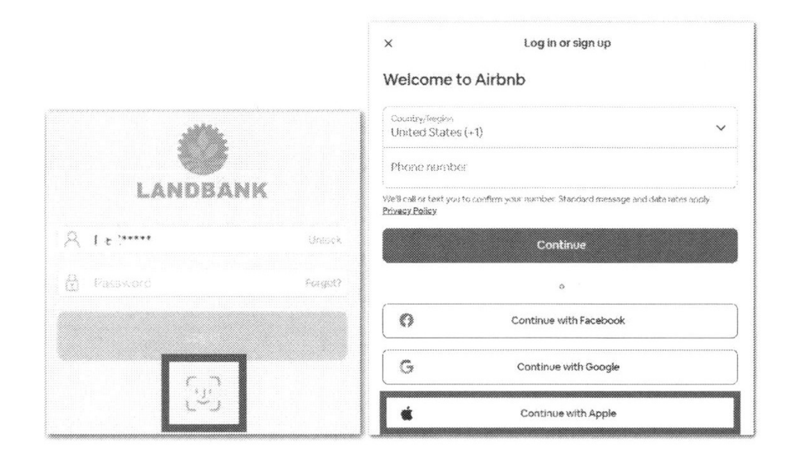

3. **Control sharing information on apps and websites.** Check the information you share by reviewing and updating it in app settings and privacy menus. If you wish to keep your information private, select the "stop sharing information" option available on the apps or websites where you registered. You may also manage app tracking on your phone. You may accept or reject permissions asked by apps or websites that want to track activities across your device. You can also change the app tracking permission in the apps or website settings where you registered.

4. **Update the iPhone system.** iOS is a mobile operating system powering Apple products, including iPhones. Apple often releases new updates on iOS to provide the best services to its users. These updates are composed of the latest features, fixes on system errors, and upgraded security measures. Therefore, it is essential to update your phone to the latest version of iOS. The data and settings are not changed when updating your device. However, it is recommended to back up your iPhone before doing so.

iCloud is a data storage software available on every Apple device. Follow these steps to back up your phone automatically using iCloud:

1. Using your old iPhone, Go to **Settings**.

2. Tap **your name** at the top of the screen. Tap **iCloud.**

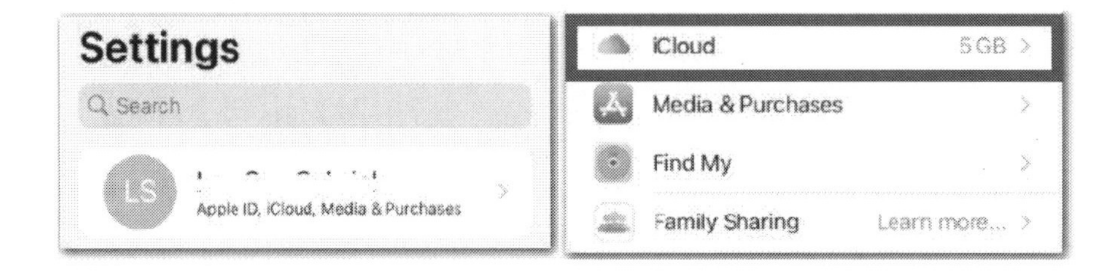

3. Select **iCloud Backup**, then switch on **Back Up This iPhone**.

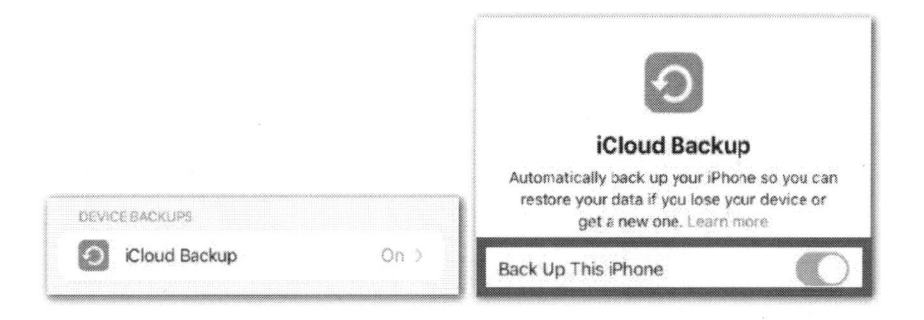

4. Select **Back Up Now**.

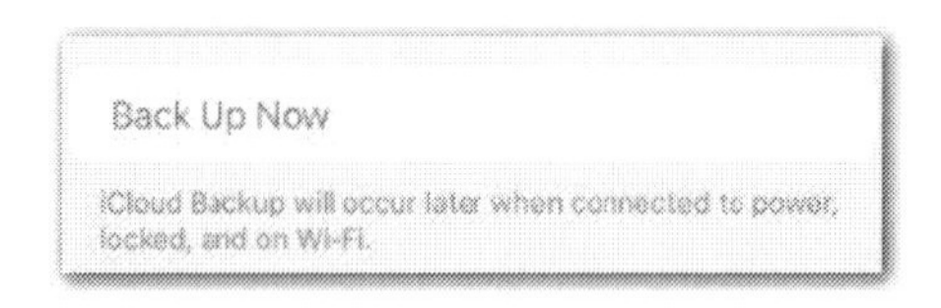

Follow the steps below on how to update your iPhone automatically and manually. **Tip**: System upda switches off your phone for a while, so if you need to use your phone immediately, consider attempting update it another time.

Automatically update iPhone

When enabled, iPhone automatically checks for updates and, if available, downloads and installs it overnight with full battery and connected to the Wi-Fi. iPhone keeps you informed about the update and its activities.

1. Go to **Settings**.

2. Scroll down or type **General** on the search bar. Tap **General**.

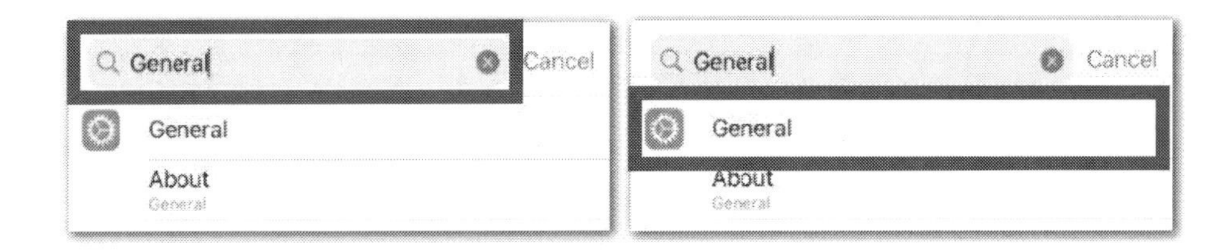

3. Tap **Software Update**, then tap **Automatic Updates**.

4. Switch on **Download iOS Updates** and **Install iOS Updates**. If you wish to turn off the automatic update, switch off **Download iOS Updates** and **Install iOS Updates**.

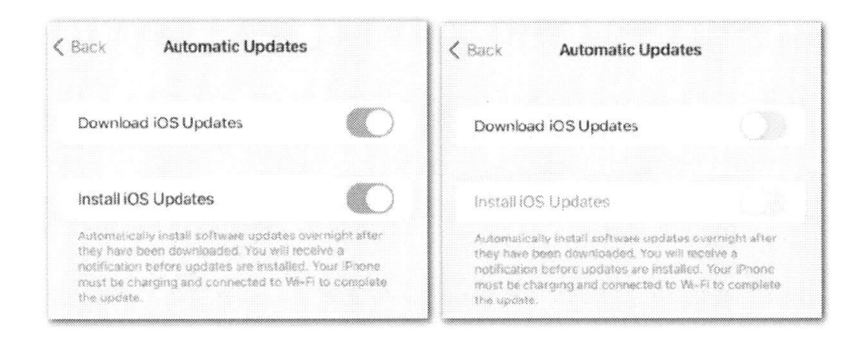

Manually update iPhone

Check if there is an update available at any time.

1. Go to **Settings**.

2. Scroll down or type **General** on the search bar. Tap **General**.

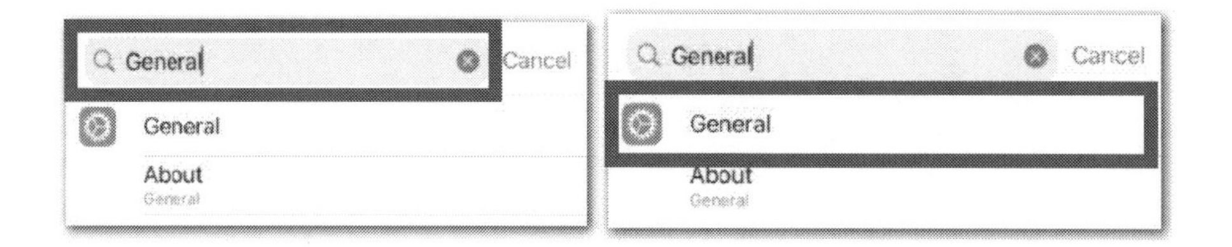

3. Tap **Software Update**. The iOS version that is currently installed is displayed on the screen as we as when an update is available. If an update is available, tap **Download and Install**.

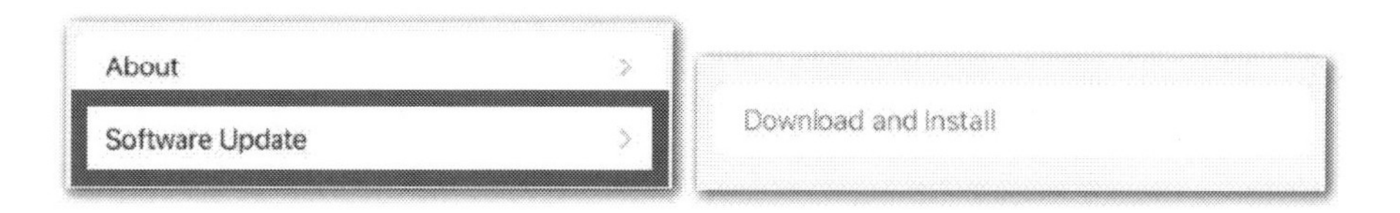

Find My

One necessary setup to optimize your iPhone is Find My. **Find My** helps locate your phone in case of lo or stolen, plus other functions. To turn on Find My:

1. Go to **Settings**.

2. Tap your name, then tap **Find My**.

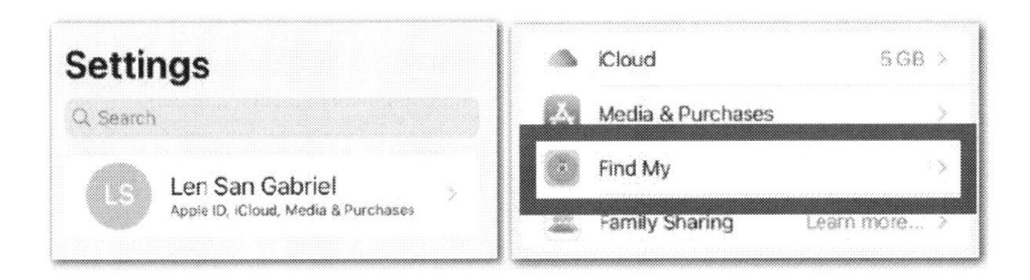

3. Tap **Find My iPhone**.

4. Switch on **Find My iPhone**.
5. Switch on **Find My network**. This will know where your device is even when it's offline, or in power reserve mode, and after powering off.
6. Switch on **Send Last Location**. When the battery is low, the location of your device will be sent to Apple.

7. If you want your family or friends to know your location, switch on **Share My Location**.

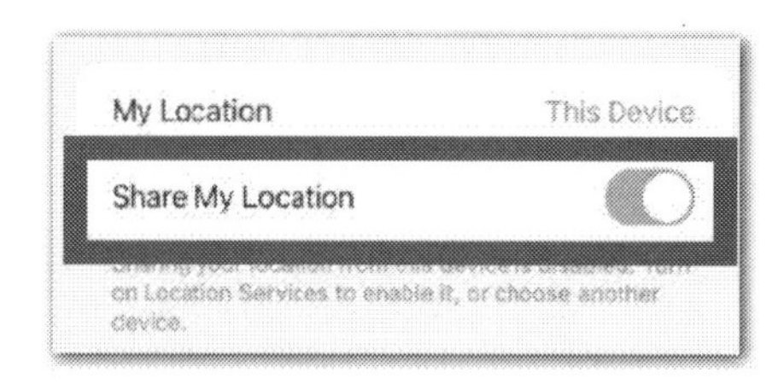

Note: Location services must be turned on to enable Share My Location.

How to Use Emergency SOS

Whether you like it or not, an emergency might happen, so it's essential to prepare for it. Emergency SOS is a feature in iPhone designed to call for help quickly in the event of an emergency. This feature benefit everyone, especially seniors and people with medical conditions. When you use Emergency SOS, th emergency authorities will get a call from your mobile. The emergency authorities receiving the emergenc call may vary depending on your location (e.g., 911 for the USA).

1. Push and hold **one of the volume buttons** with the **side button** until the Emergency SOS slide shows.

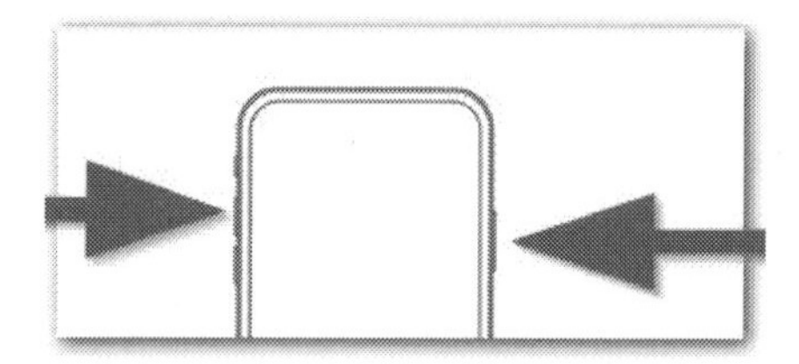

2. Slide the **Emergency Call slider**. Note: Long pushing the side and volume buttons will show a countdow and ring an emergency alert sound. If you continue to hold down the buttons until the countdown stop your emergency call proceeds.

How to Use Medical ID

Use Medical ID with Emergency SOS. Medical ID is your health information accessible on your Lock Screen after you call for an Emergency SOS. This helps the first responders to learn your basic medical record, which is crucial for treatment.

Set up Medical ID

1. Go to **Health**.

2. Tap the **Summary** tab at the bottom, then tap your **profile picture** at the top-right corner of your screen.

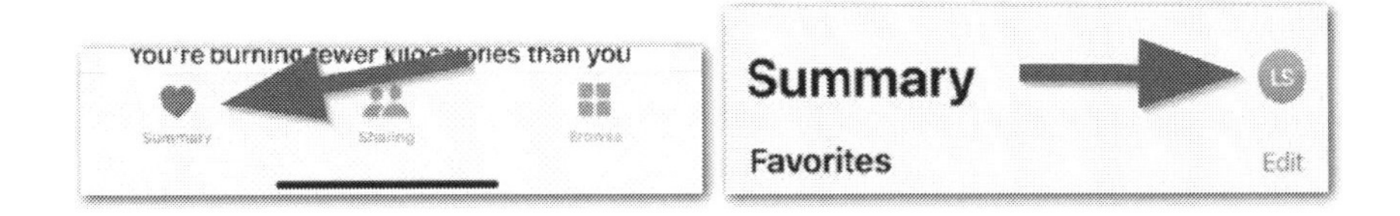

3. Tap **Medical ID** under your profile picture, then tap **Edit** at the top-right corner of your screen.

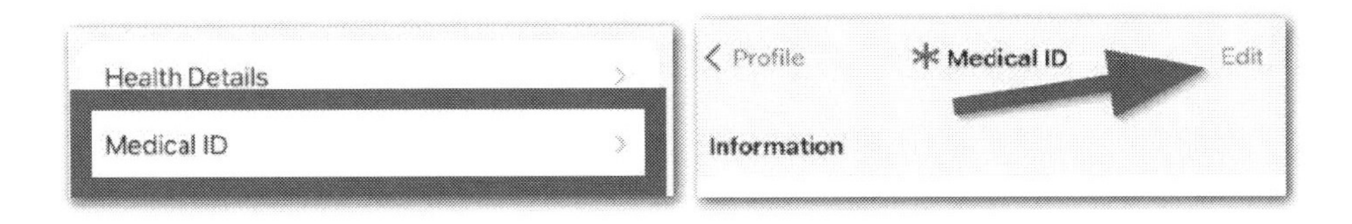

4. Enter your health information.

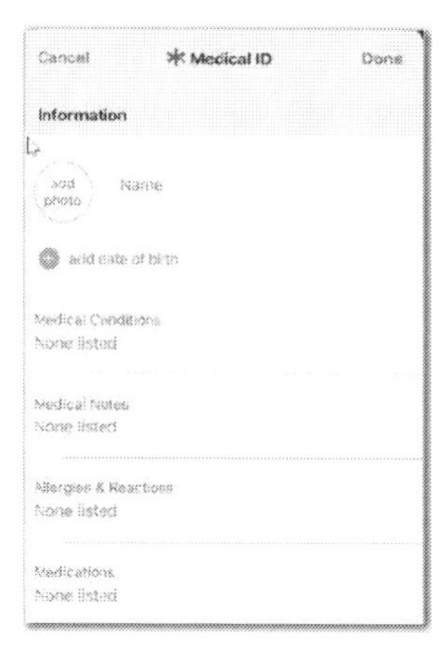

5. To add emergency contacts, tap the **add button** near the «add emergency contact.»

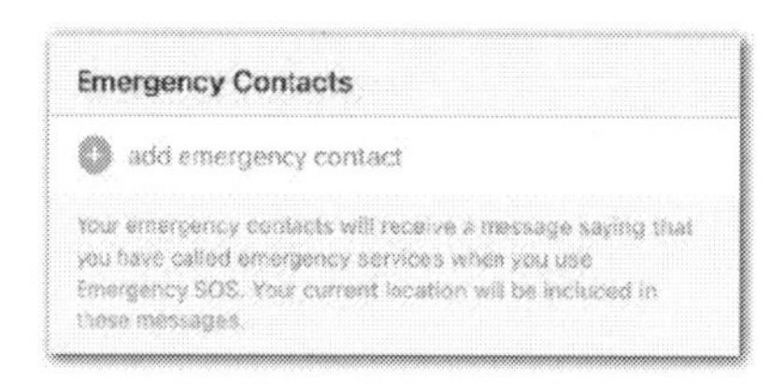

6. Tap a contact, then select their relationship.

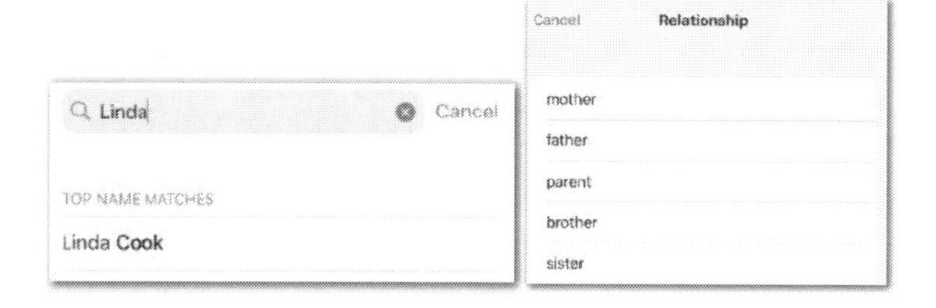

7. Switch on **Show When Locked**. This will make your Medical ID accessible from the Lock screen an give emergency responders information.
8. Switch on **Share During Emergency Call** to automatically share your medical ID with the emergenc services during the emergency call (applicable only in USA and Canada).

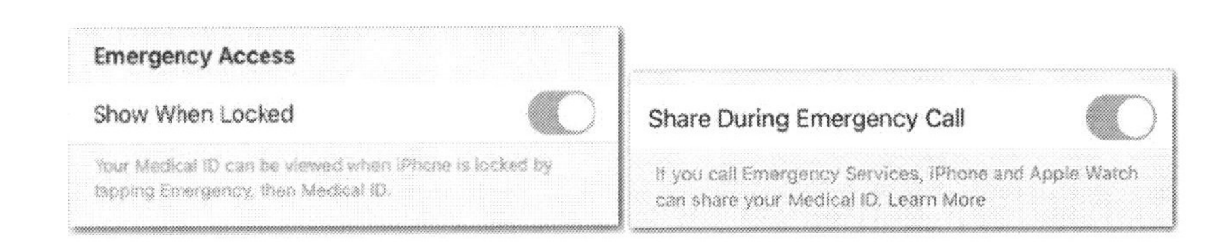

9. Tap **Done** at the top-right corner of your screen.

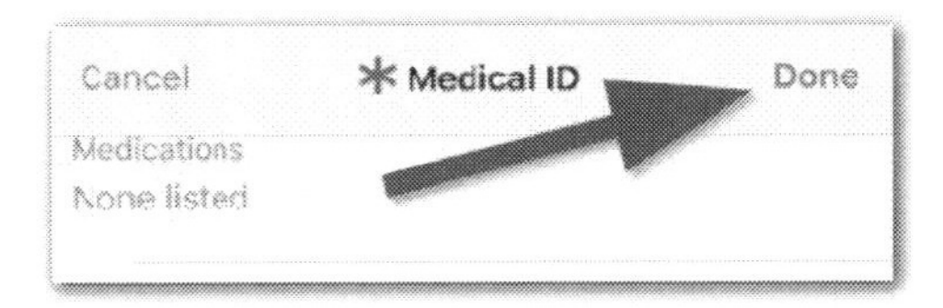

CHAPTER 09

Extras

At this point, you have learned about the basics of your iPhone. Next, you will be informed of some additiona functionalities on your device that can help you live a convenient lifestyle. You might find Apple apps an services like Wallet and Family Sharing helpful. The Wallet app lets you organize cards, tickets, and more It also introduced the Apple Pay service for online and offline payments. With Family Sharing, you ca access and share various Apple services with family members. Another feature covered in this chapte is the advantages of syncing the iPhone with the family of Apple devices, plus the guide on how to sync.

What is Apple Wallet

Apple Wallet lets you safely and easily save your transit passes, boarding passes, tickets, rewards card and more. This app organizes your amenities and imports them to your phone - all in one place, so yo don't have to keep looking for missing tickets and cards. Apple Wallet uses your Apple ID to store yo card information securely, making it easy to add, edit, and manage your cards and passes across all yo devices. Apple Wallet also features Apple Pay, which you can use to make purchases using iPhone.

What is Apple Pay

Apple Pay is a mobile payment service developed by Apple for secure, contactless payments online ar offline. It enables quick and convenient transactions at many outlets supporting Apple Pay while using yo device.

Set up Apple Pay

Add a debit, credit, or prepaid card to the Apple Wallet

1. Go to **Wallet**.

2. Tap **Add**. Then, tap **Debit or Credit Card** to add a new card.

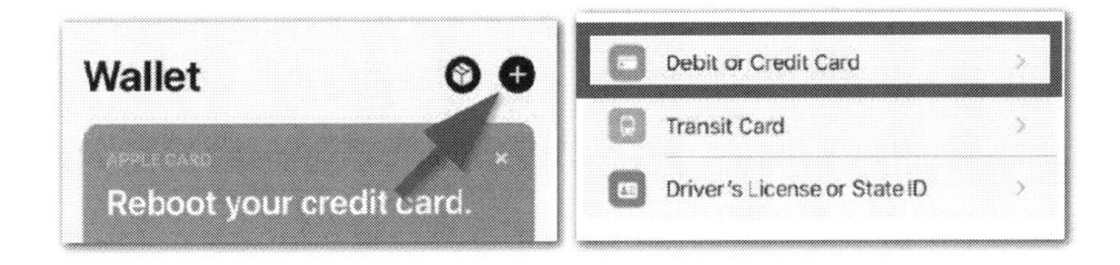

3. Tap **Continue**. Follow the prompts on the screen to add a new card.

4. Verify your information with your bank or card issuer.

Note: Only the supported cards can use Apple Pay. You may check with your card issuer if Apple Pay supports your cards.

Remove a card as a payment method

Tap the card you want to remove, then tap the **Remove Card**.

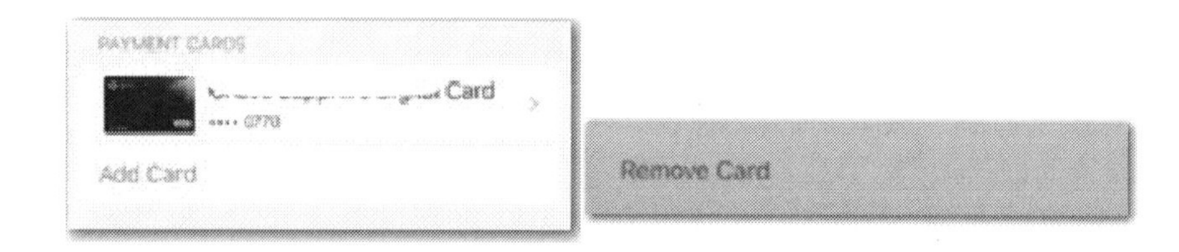

2. Tap **Remove**. Your card will be added to the "Previous Card menu."

Add a boarding pass, ticket, or other passes to Wallet

1. Locate your pass from apps or email, then tap **Add to Apple Pay**. Follow the prompts on the screen

Use a boarding pass, ticket, or other passes in Apple Wallet

1. To open your pass, tap a **notification** if there's any. If none, double-tap the **side button** (For iPhone 13 and 14) or double-tap the **Home button** (iPhone SE). Proceed with the security details whe asked.

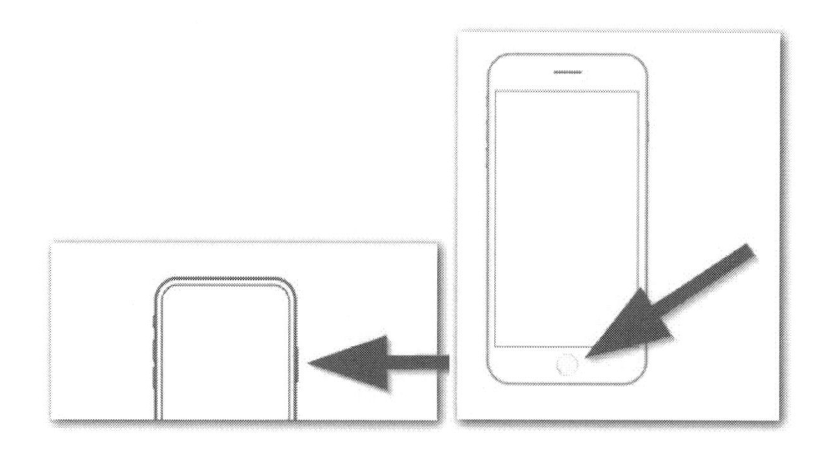

2. Tap your **default payment card** to see your other cards. Then, tap your **passes**. Proceed with the security details when asked.

3. Present it to the attendant or scan your **iPhone** at the boarding pass, ticket, or other passes' reader.

What is Family Sharing and How to Set it Up

amily sharing allows up to five family members to share subscriptions to Apple services without knowing each ther's Apple ID. Share Apple exclusive offers with your family, like Apple Music, Apple TV+, Apple News+, pple Arcade, Apple Books, and App Store purchases. You can also share an iCloud storage plan, a family hoto album, or help locate a missing Apple device.

et up a family group.

om the iPhone, the family organizer (must be an adult) can set up Family Sharing for the entire group.

. Go to **Settings**.

2. Tap your name, then tap **Family Sharing**. Tap **Continue**.

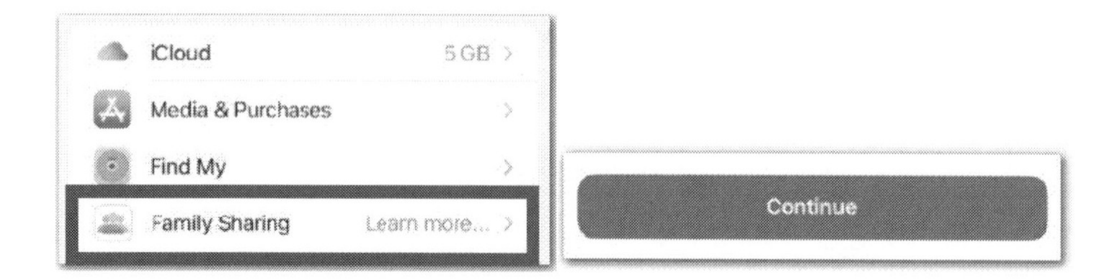

3. Follow the prompts on the screen.

Invite people to join your family

You can send an invitation to join your family via message, email, or in person.

1. Go to **Settings**.

2. Tap **Invite** to invite a family from the suggestions.

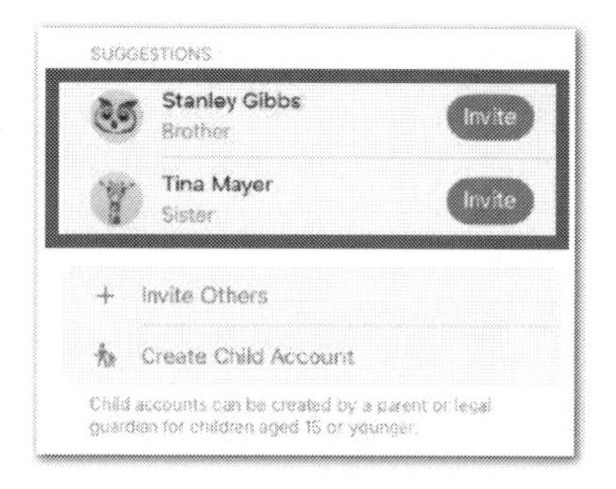

3. Tap **Create an Account for a Child** if your child doesn't have an Apple ID. Otherwise, tap **Invit People**.

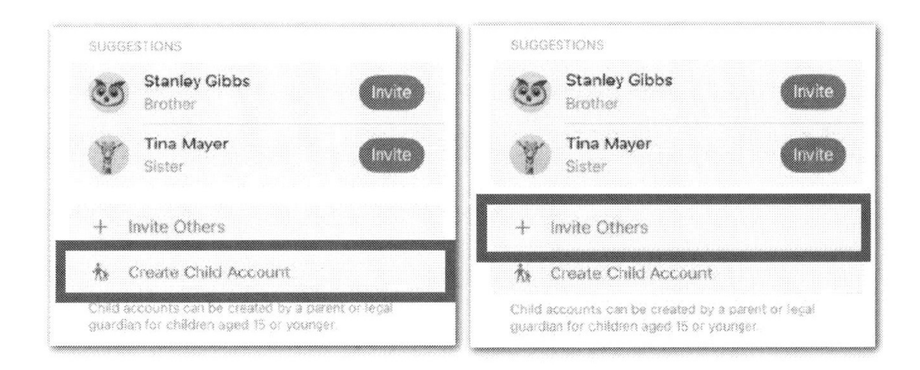

4. Follow the prompts on the screen to set up family-sharing features like controls, location sharing, and more.
5. Those who received the invitation can join directly from a text message, email, or in person.

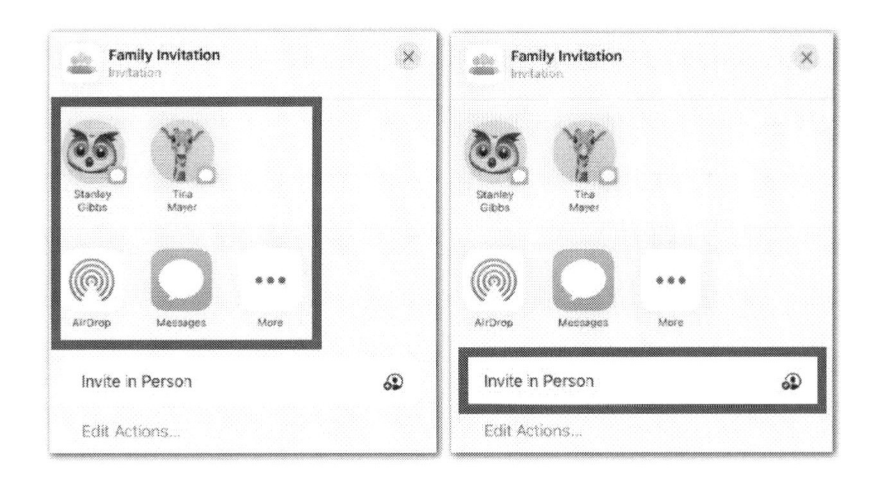

How to Sync Your iPhone With Your Mac

hen you sync your iPhone with other Apple devices, such as Mac, you can transfer contents and keep) to date between all other devices. So, the next time you want to view photos taken from your iPhone, u can view them on your Mac, as they also show on this device.

syncing, you can safely save a copy of your device's data to your Mac and use it as a backup to restore st data. You can also automatically sync all files between devices for easy updating or select only the es you want to sync.

Sync iPhone with Mac

1. Using a USB or USB-C cable, connect your device to your Mac.
2. On your Mac, click **Finder**, then select your iPhone in the Finder side navigation bar.

3. Select the content type you want to sync in the top menu bar.

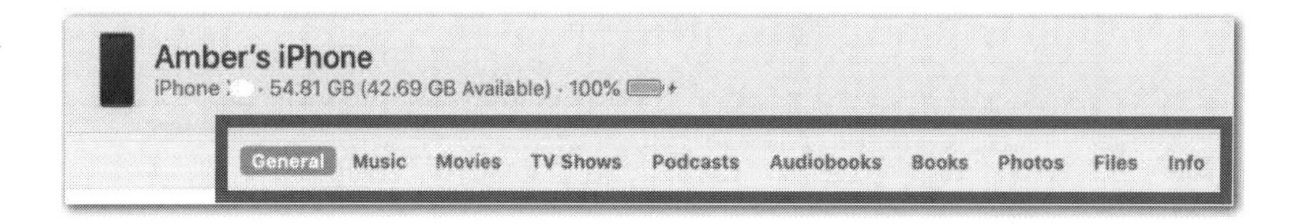

4. Tick the **"Sync (content type) onto (iPhone name)"** box. Repeat steps 3 and 4 for all items you wis to sync.

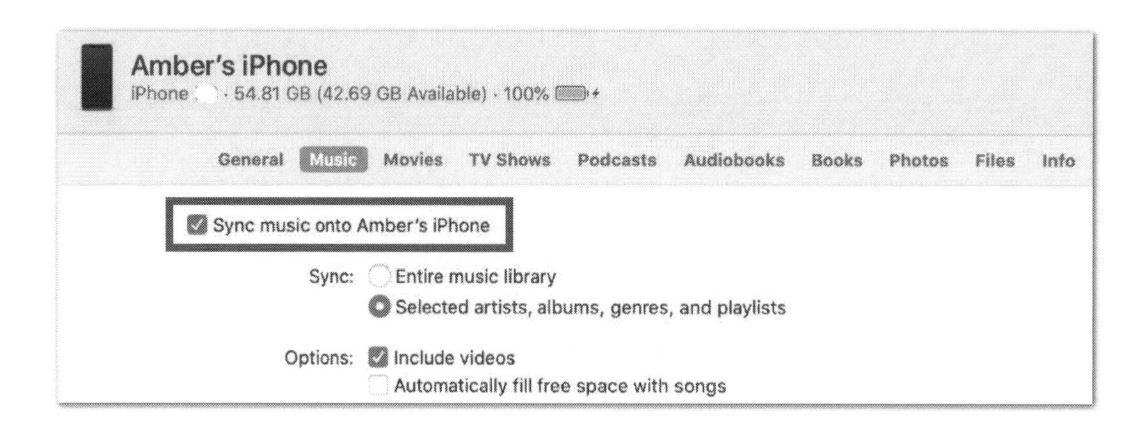

5. Click **Apply**.

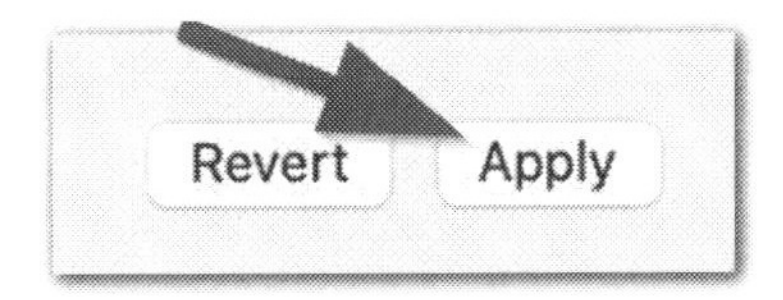

Automatically sync contents between devices

1. Click **General** in the top menu bar.

2. Select the **"Automatically sync when this (iPhone) is connected"** box.

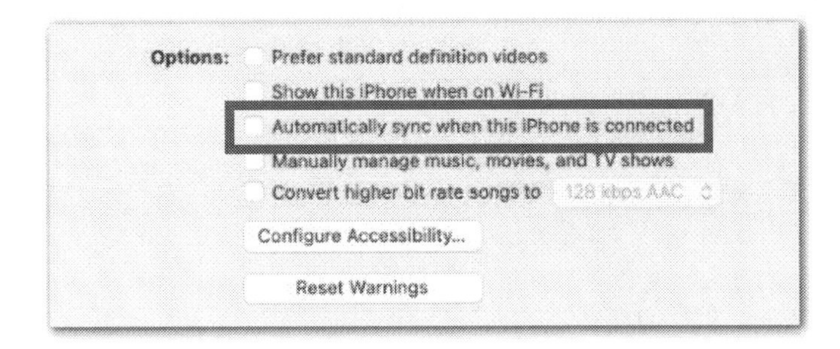

3. Turn the switch on, then select sync settings.

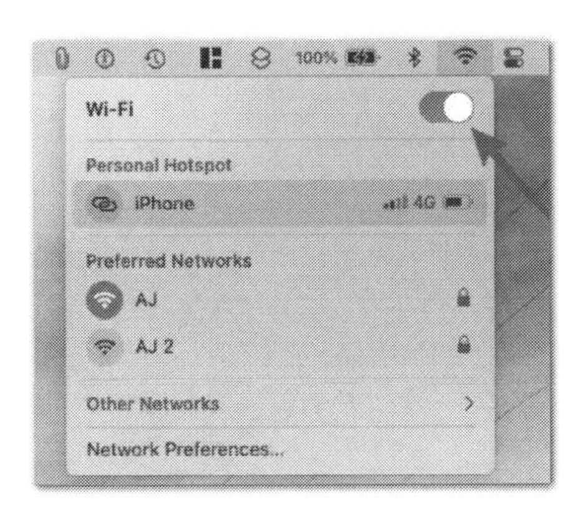

4. Click **Apply**.

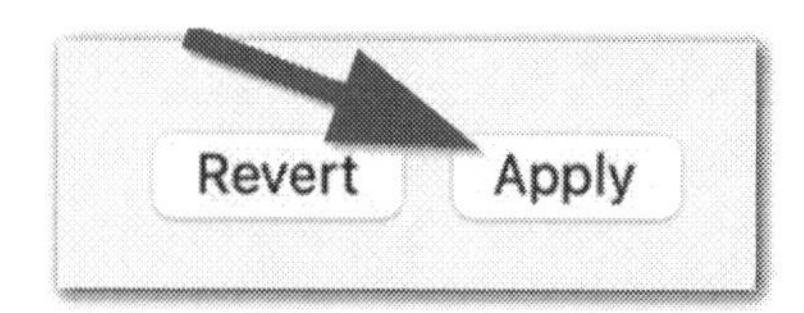

You can set up syncing over Wi-Fi, so you can sync the next time wirelessly. This setup still requires a connection between iPhone and Mac with a USB or USB-C cable. After you switch on the option to sync over Wi-Fi, you can sync without the cable the next time you sync your devices.

Note: iPhone and Mac must be connected to the same Wi-Fi network to sync wirelessly.

1. Click **General** in the top menu bar.

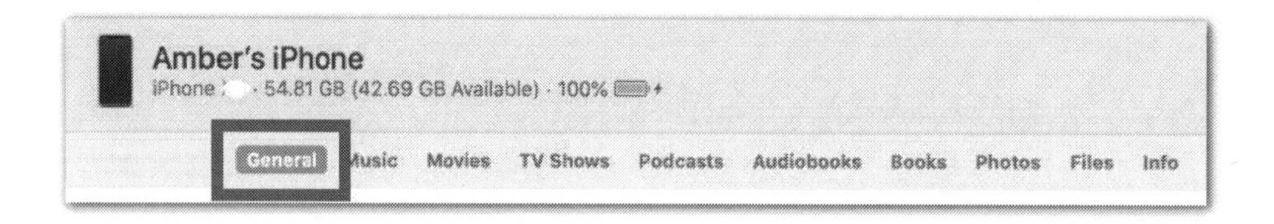

2. Select the "Show this **iPhone** when on Wi-Fi" box.

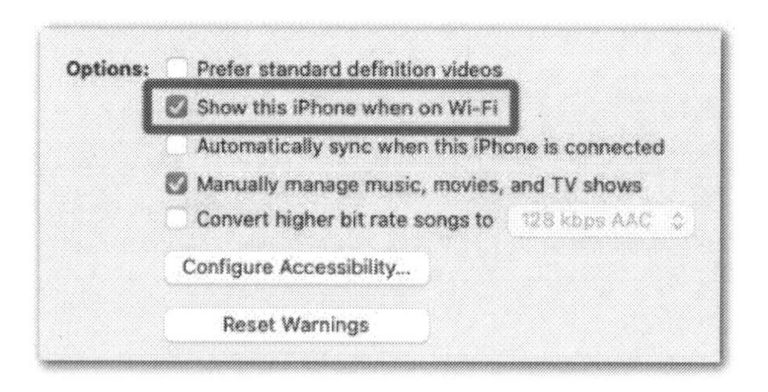

3. Turn the switch on, then select sync settings.

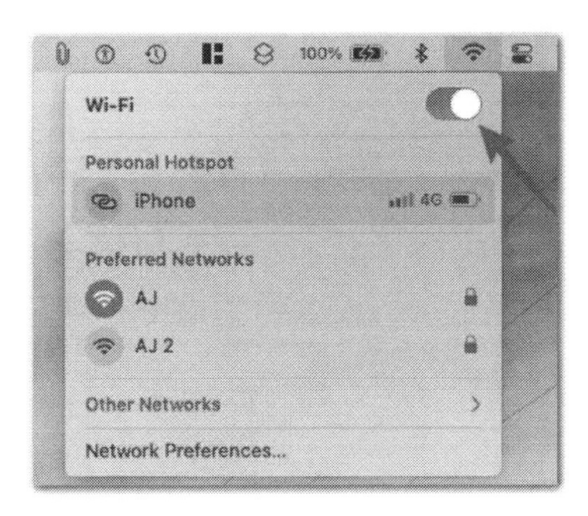

4. Click **Apply**.

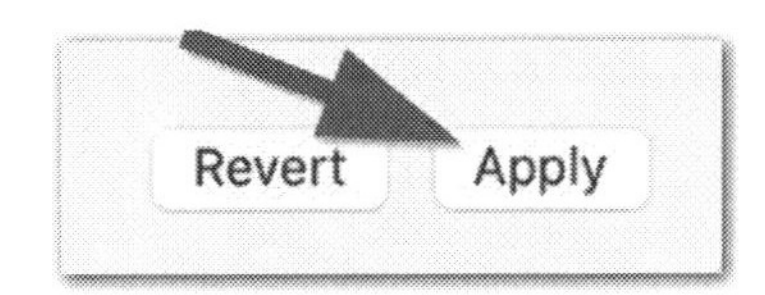

5. Click the **eject button** in the Finder sidebar to disconnect your device.

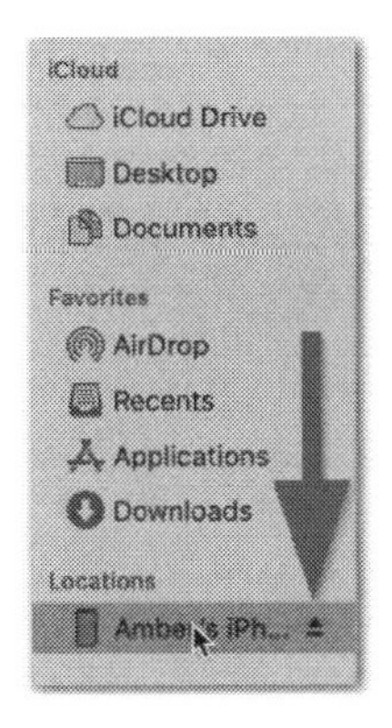

CHAPTER 10

FAQ

Got any questions about your iPhone? It's quite understandable when, at times, you feel stuck using you smartphone — and we can't blame you! iPhone is crafted with numerous robust functionalities, multipl navigations, and shortcuts to give you the best phone experience. Fortunately, this section provide answers to your Frequently-Asked-Questions (FAQs) that will make you love your device even more.

How Can I Get Rid of Notifications

Notifications can sometimes get in the way when you wish to navigate a specific feature on your phon or view the screen fully. It may also frustrate you when your notification tone rings continuously during group chat party. To avoid these situations, you can dismiss, remove, or mute your phone notification i the following ways:

Dismiss a notification

1. While using another app, view a notification you receive by tapping on it. Dismiss by swiping up.

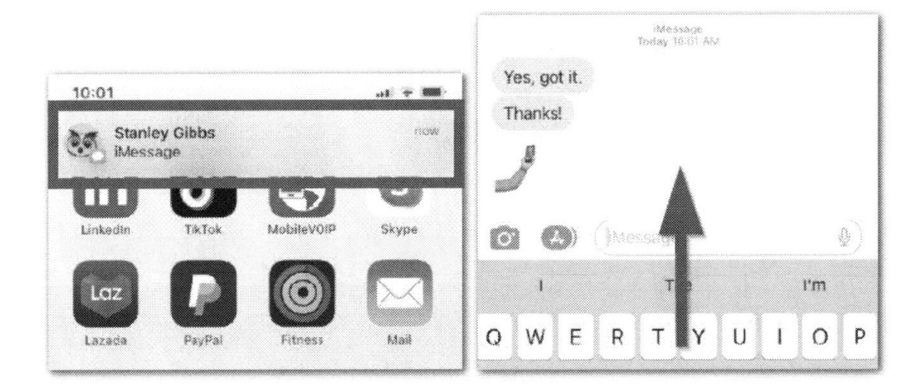

Clear a notification

1. To clear all notifications from the lock screen, swipe up on the lock screen. Then, swipe left on a notification or group of notifications, then tap **Clear** or **Clear All**.
2. To clear all notifications in the Notifications Center, swipe down from the top-left corner of your screen to go to the Notifications Center. Tap **Clear** or **Clear All**.

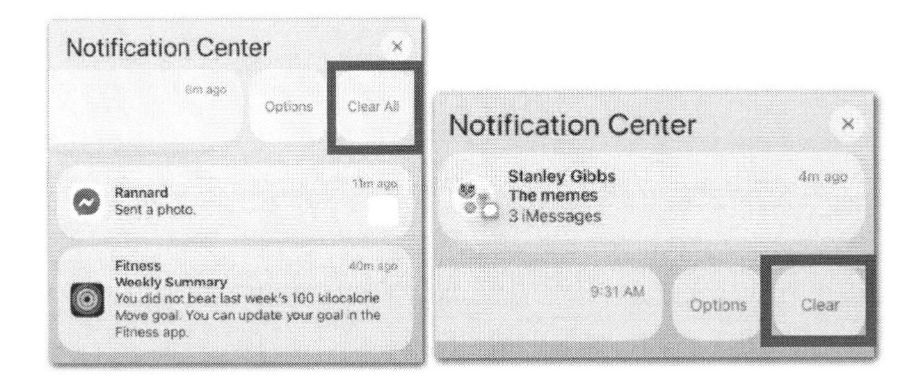

Mute notifications for an app:

Swipe left on the notification or group of notifications, then tap **Options**. Tap mute to silence the notifications from the app for an hour or a day.

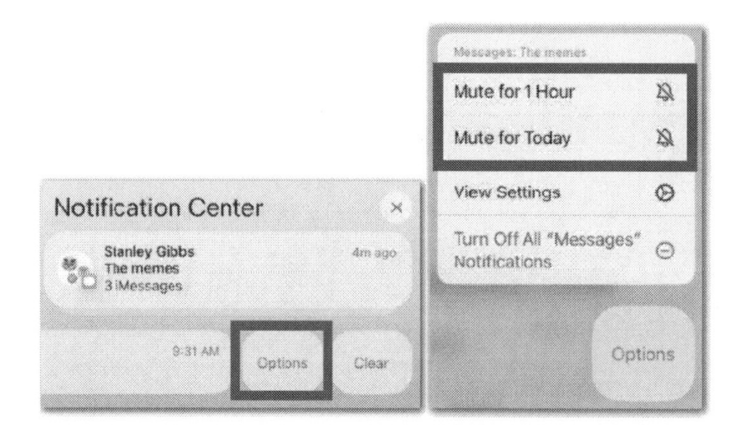

2. To **mute all notifications, go to Control Center.** To view the Control Center, swipe down from the top-right corner of your screen (for iPhones 13 and 14) or swipe up from the bottom (for iPhone SE).

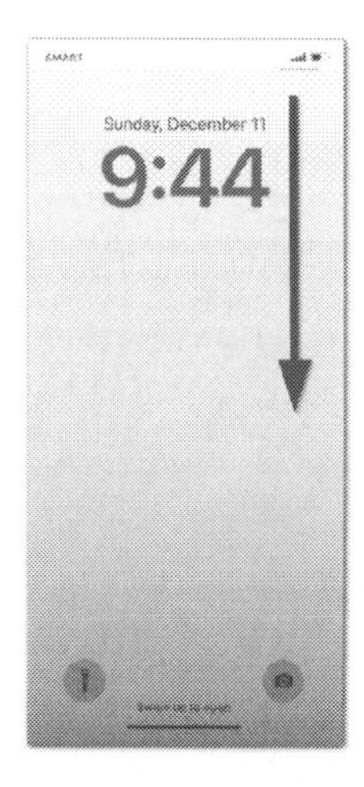

3. Push and hold the moon icon. Tap **Do Not Disturb**.

How Can I Check My Screen Time

With Screen Time, your phone records how much time you spend using your device, which apps you spend a long time on, and more. This information will help you understand and manage your screen activities. You must first turn on Screen Time to check how long you spend on your device.

Turn on Screen Time

1. Go to **Settings**.

2. Scroll down or type **Screen Time** on the search bar. Tap **Screen Time**.

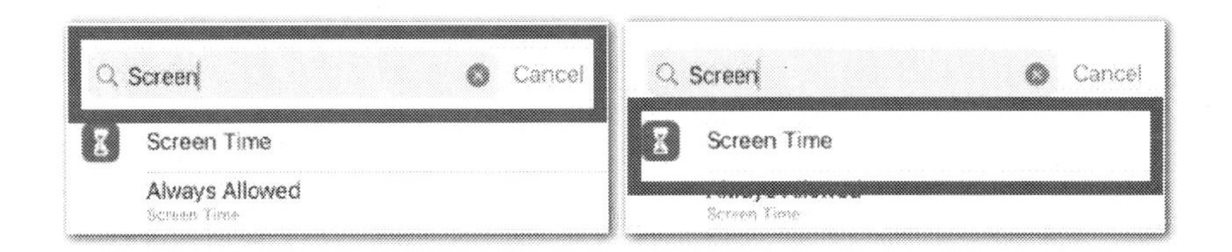

3. Tap **Turn On Screen Time**, then tap **Turn On Screen Time**.

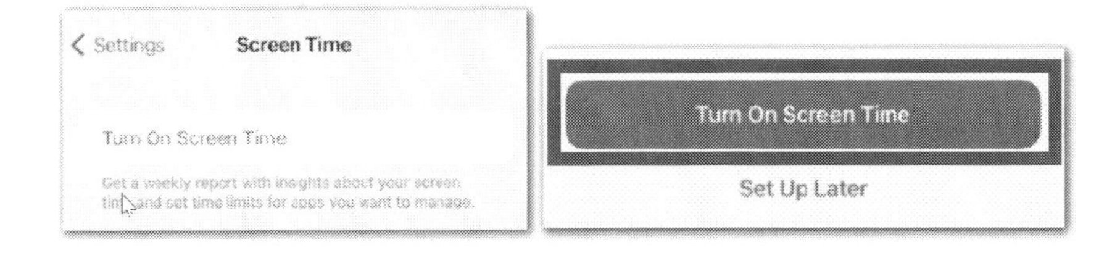

4. Tap **This is My iPhone** to set up Screen Time for yourself.

Check your Screen Time

1. On Screen Time, tap **See All Activity**. Select between **Week** and **Day** to see your Screen Time summary.

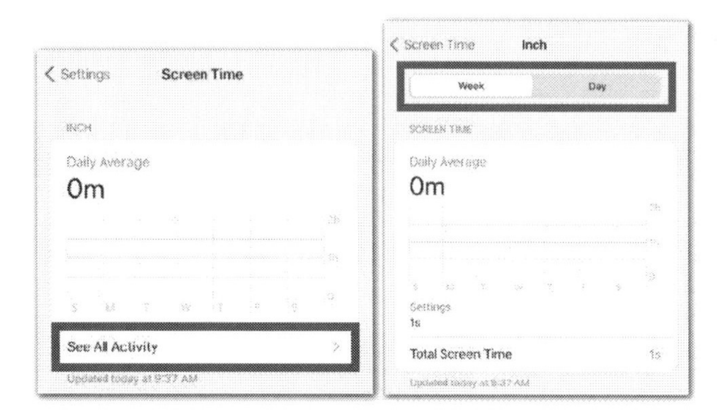

How Can I Take a Screenshot

On your iPhone, quickly capture a screenshot of what is on your screen.

Take a screenshot with iPhones 13 and 14

1. Quickly push the **side button** and the **volume up button** at the same time.

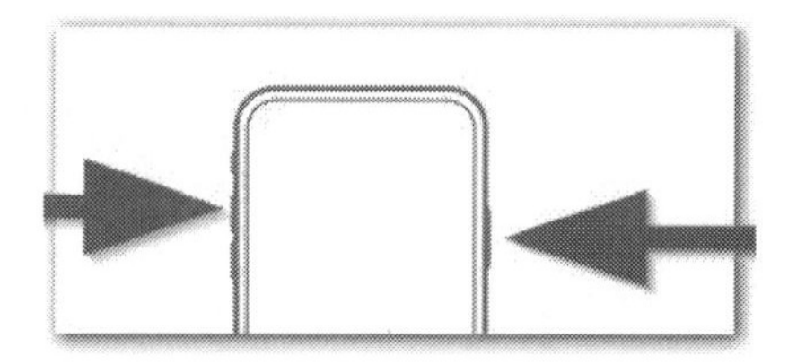

Take a screenshot with iPhones SE

1. Quickly push the **side button** and the **Home button** at the same time.

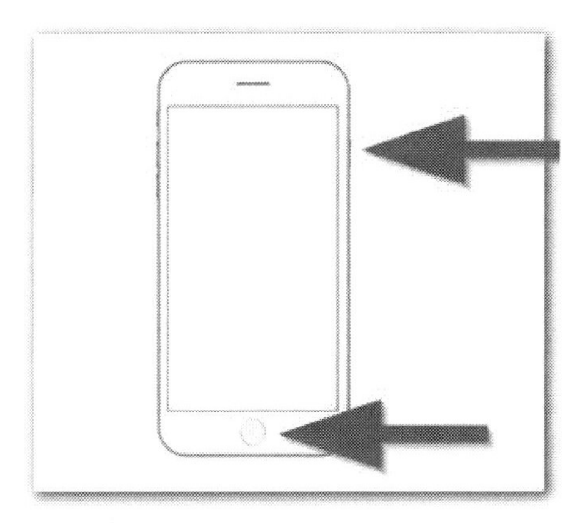

2. A successful screenshot will temporarily show a thumbnail in the lower-left corner of your screen. To open the thumbnail, tap it or swipe left to dismiss. View your screenshot in **Photos**.

How to Enable Location Services

witching on location services on iPhone enables apps and websites to give you more accurate information d related content you require of them. You can choose how you want to share your location by using cation services. Among them is the option to share it to specific apps only, use it once, or always when s used.

To enable location services:

1. Go to **Settings**

2. Scroll down or type **Privacy & Security** on the search bar. Tap **Privacy & Security**.

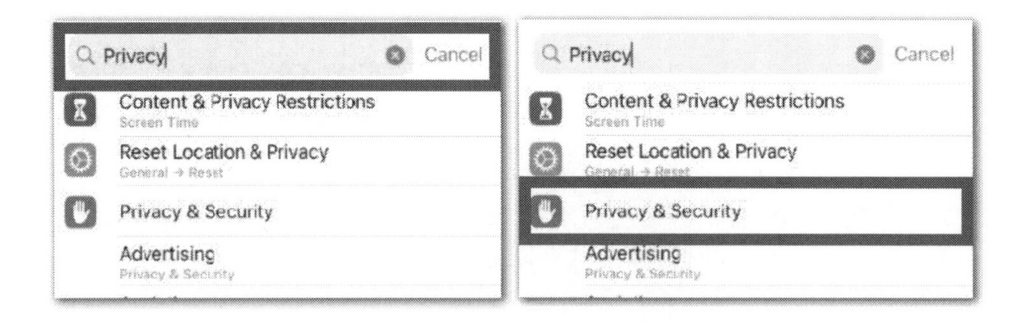

3. Switch on **Locations Services**.

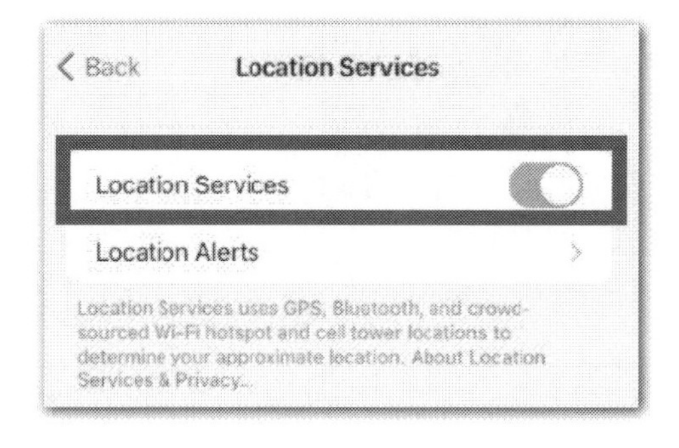

How to allow an app access your location

Some apps and websites can only give you better services with location services on your phone turned o
Depending on which data the apps would like to access, you will receive a notification like the one belo
asking for permission to access your location. You can select any of the following:

- Only While Using the App
- Always Allow
- Don't Allow

How to turn Location Services on or off for a specific app

1. Go to **Settings**

2. Scroll down or type **Privacy & Security** on the search bar. Tap **Privacy & Security**.

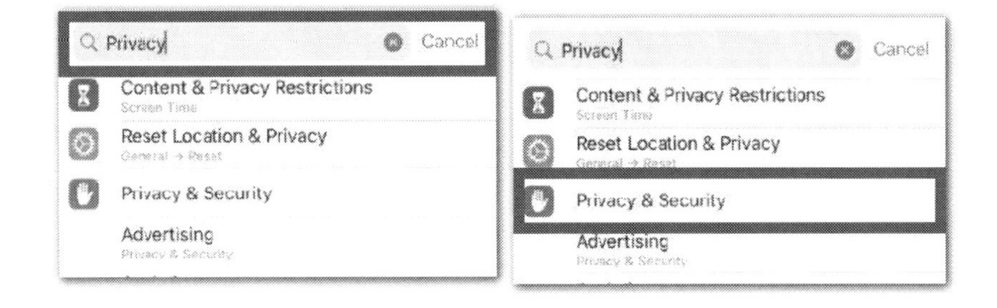

3. Tap **Locations Services**, then switch on **Location Services**. Scroll down to find the app.

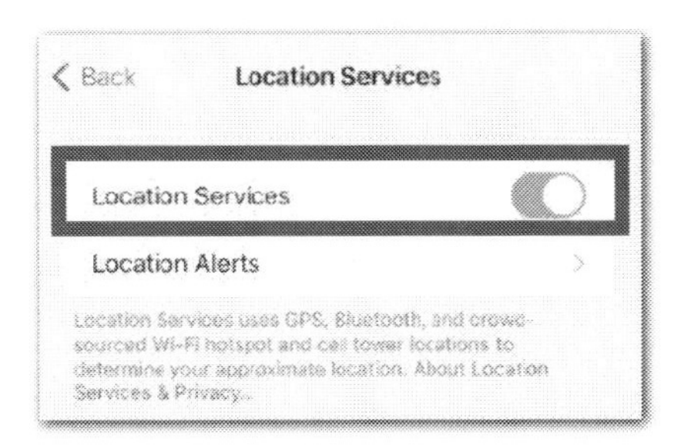

4. Select the app, then select any of the following options under Allow Location Access:

- **Never** - Disapproves access to Location Services.
- **Ask Next Time or When I Share** – This will prompt an app to ask permission every time it needs to access your location or only when you wish to share it.
- **While Using the App** – Allows access to Location Services only when the app or one of its features is visible on screen.

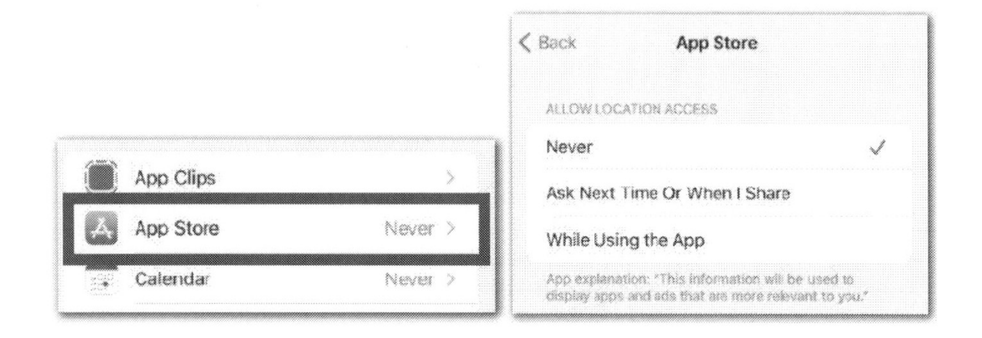

CHAPTER 11

TROUBLESHOOTING

No matter how intuitive and easy to use the iPhone, it can't always run smoothly as you wish. Bumping into problems on your device is common, but don't let it ruin your day. Look up for solutions and be patient in troubleshooting these issues. The good news is that this guide is packed with easy tutorials on fixing some of the most common problems encountered on iPhones, so you don't have to do your research and get overwhelmed with too much information.

What If You Forget Your Password or PIN Code

After too many times of entering the wrong passcode on your iPhone, a prompt will inform you that your iPhone is disabled. If you forgot your iPhone passcode, the best way to fix this is to put your device into recovery mode using a computer. This means the data on your phone will get erased, so you can set it up again. After resetting, you can restore your data and settings from a backup. You can refer to Chapter Two: Setting up your iPhone on the page 6 to help you set up and transfer data.

Prepare these tools:

- A Mac or PC (Windows 10 or later with iTunes installed)
- USB cable, USB-C cable, or any compatible cable to connect your iPhone to the computer.

If a computer is not available to you, go to an Apple Retail Store or Apple Authorized Service Provider for assistance.

Put your iPhone into recovery mode

1. Push both the **side button** and the **volume down button** until the power off slider shows. Move the **power off slider** to turn off your iPhone.

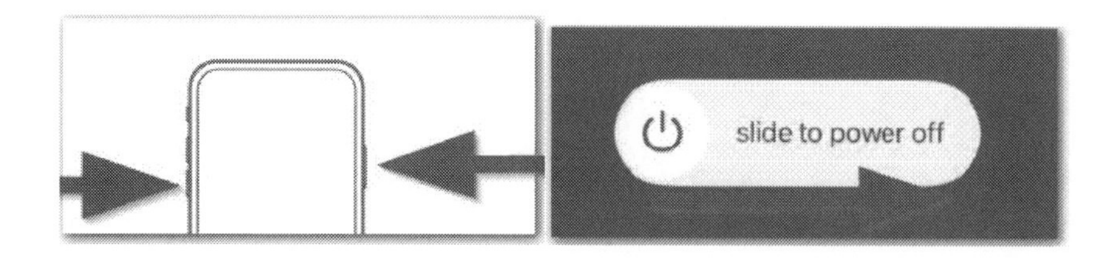

2. Push the **side button** while quickly connecting your iPhone to the computer. **Hold the button and don't let go** until the recovery mode (image where a cable must be connected to a computer) shows on your iPhone screen. If you see the passcode screen, turn off your iPhone and start the recovery mode again.

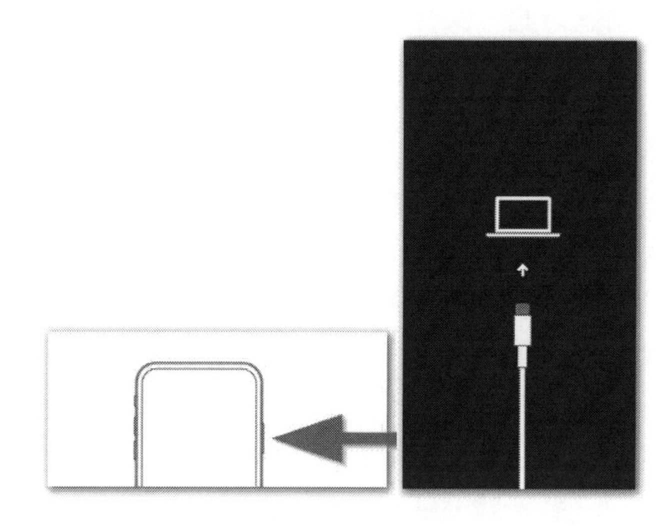

3. Find your iPhone in iTunes (PC) or the Finder (Mac). When the Restore or Update option show select **Restore**. Your computer starts the recovery procedure by downloading software for your iPhone. Let the download complete, if it takes more than 15 minutes and your device leaves the recovery mode screen, turn off your iPhone and try again.

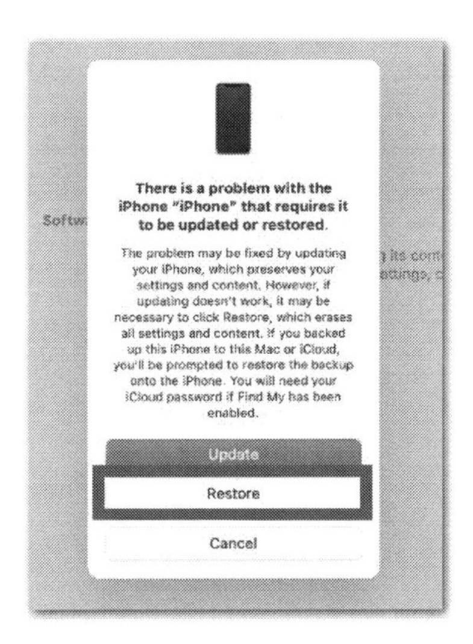

4. Finish the process, then disconnect your iPhone from the computer. You can refer to Chapter Two: Setting up your iPhone on the page 6 to help you set up and transfer data.

Vhat Happens If Your hone Turns Black or Freezes

you experience your iPhone that turned black, freezing, or not responding when touched, this is what ou can do:

. Quickly push the **volume up button**.

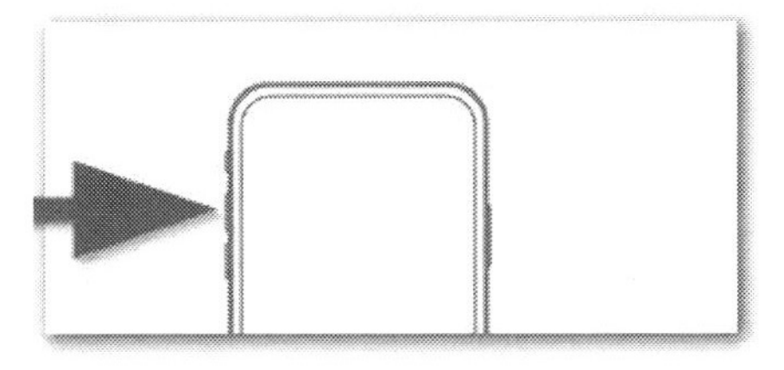

. Quickly push the **volume down button**.

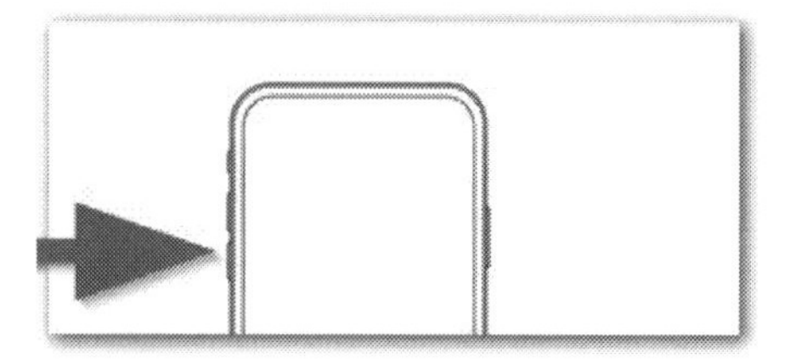

3. Push and hold the **side button** to turn it on or until the apple logo shows.

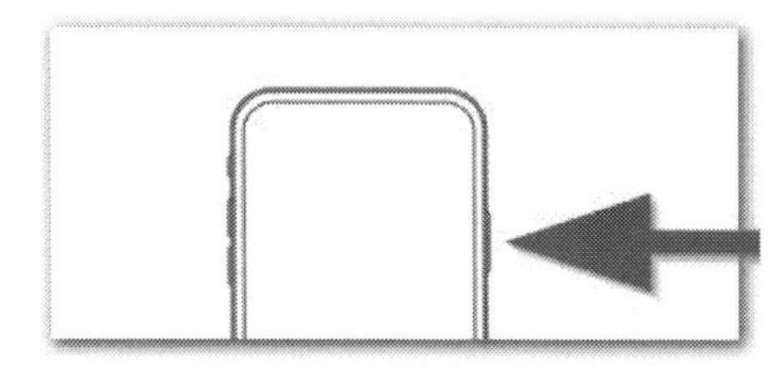

If your device doesn't respond to any of these steps, try charging it for half an hour. If it doesn't work, check if the issue is caused by charging.

Why is My iPhone Not Charging When Plugged In

Try out the steps below to check if your phone charges properly.

1. Check if your charging accessories are damaged. A damaged cable or adapter can interrupt charging so consider replacing them.
2. Try a different outlet or use a wall power outlet to check your charging cable, USB wall adapter, an wall outlet or AC power cable connections.
3. Clean the charging port on the bottom of your device and securely connect your charging cable.

Charge your device for half an hour, then push and hold the **side button** to turn it on or until the apple log shows.

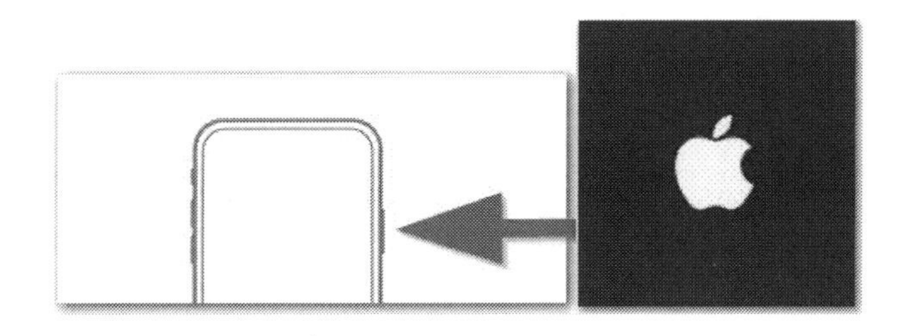

'hat If an App Freezes, But Everything Else is ,kay

f you're experiencing app crashes or frequent app freezes, this may be because the current version of the ıpp needs an update or the latest iOS is not compatible with the to run the app. Lack of storage may also ıe a cause for an app to freeze. Generally, if the device is overloaded with data, such as media and apps, :s overall performance is affected as there isn't enough space for the app to execute work as expected.

ollow any of these several troubleshooting techniques to save your app from crashing. If a step didn't 'ork, try the other approaches until your app works fine.

tep 1: Restart iPhone

. Push both the **side button** and the **volume down button** until the power off slider shows. Move the **power off slider** to turn off your iPhone.

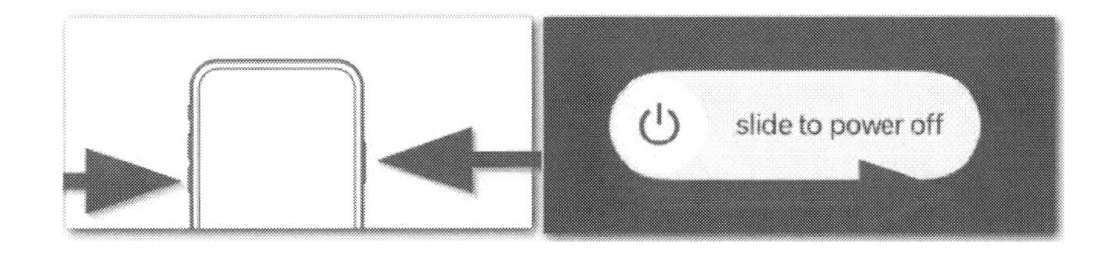

. Push and hold the **side button** to turn it on or until the apple logo shows, then try using the app again.

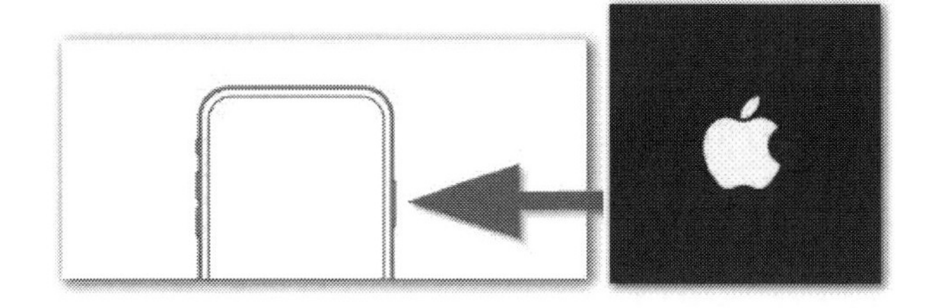

Step 2: Close and re-launch app

For iPhone 13 and 14

1. Tap **Assistive Touch**, then double tap the **Home button**.

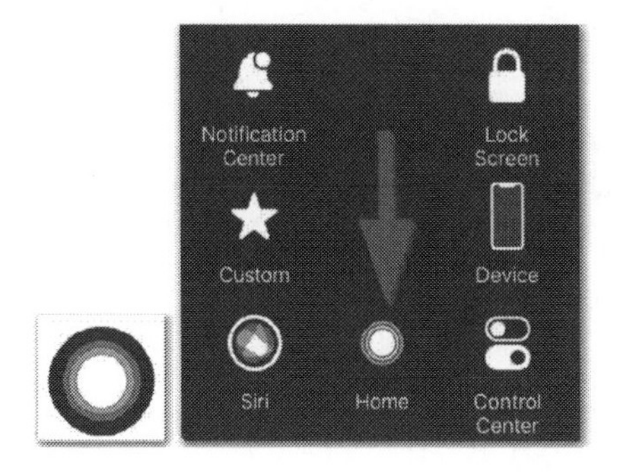

2. Swipe up the app to quit.

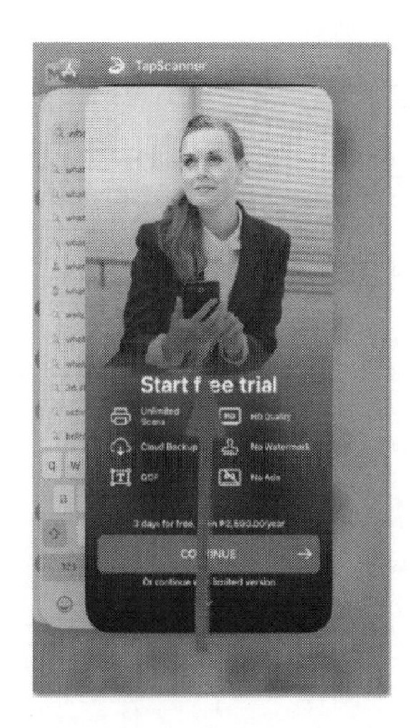

3. Tap **AssistiveTouch**, then tap the **Home button** to return to the Home Screen.

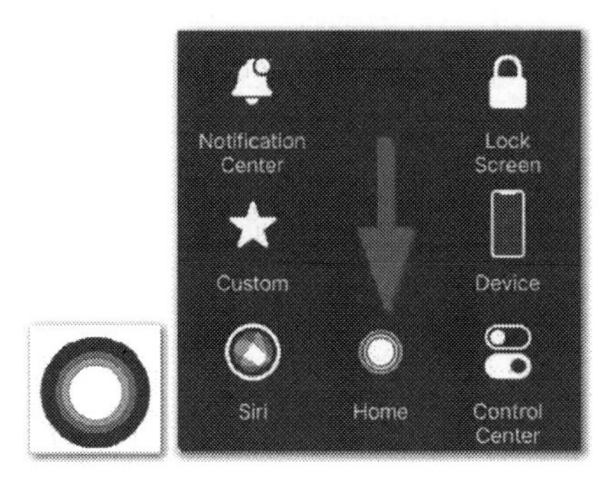

4. Tap the app to relaunch.

or iPhone SE

1. Double tap on the **Home button**.

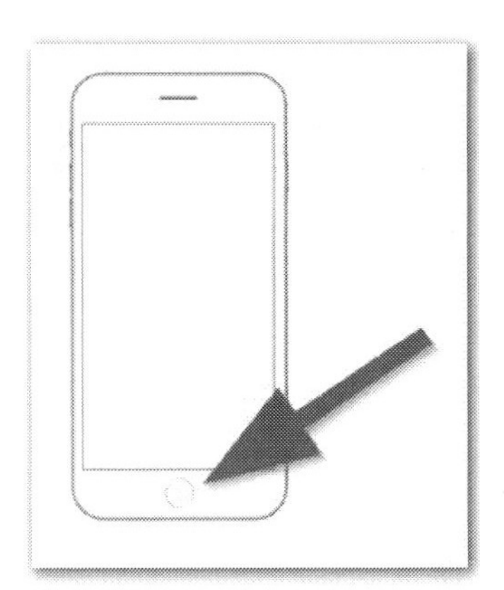

. Swipe up the app to quit. Tap the **Home button** to return to the Home Screen.

3. Tap the app to relaunch.

Step 3: Check for updates

Update the App

1. Go to **App Store**.

2. Tap **Search** at the bottom of your screen, then type the name of the crashing app on the search ba

3. Tap the app name from the search result, then tap **Update**. After the update, try using the app agai

Jpdate iOS

Jpdating iPhone's operating system, iOS, brings the latest features, fixes on system errors, and upgraded security measures, so users can smoothly navigate across apps. You may want to update iOS to its latest ersion to repair any possible bugs that may cause app crashes.

Before updating iPhone, it is recommended to do a backup. To back up your phone automatically using Cloud:

1. Go to **Settings**.

2. Tap **your name** at the top of the screen. Tap **iCloud.**

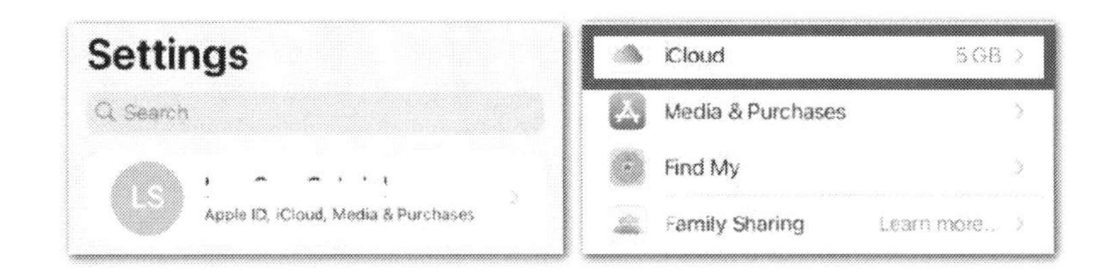

3. Select **iCloud Backup**, then switch on **Back Up This iPhone**.

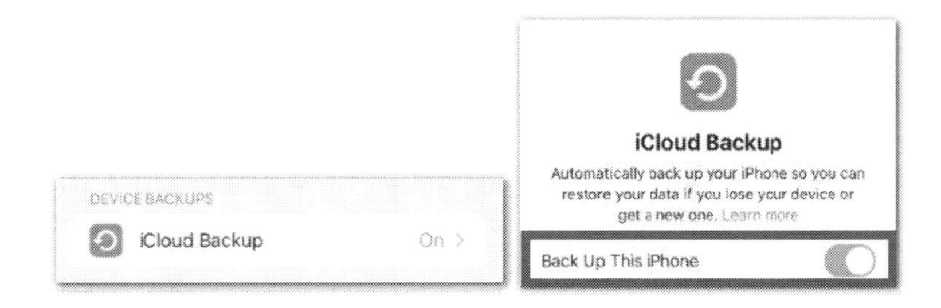

4. Select **Back Up Now**.

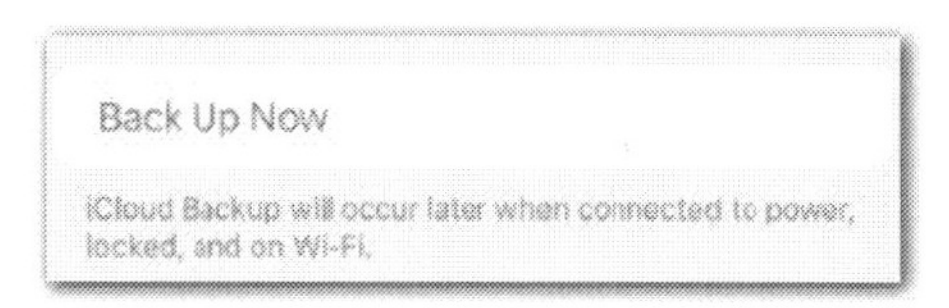

After the backup is finished, follow the steps below on how to update your iPhone manually. **Tip**: System update switches your phone for a while, so if you need to use your phone immediately, consider attempting to update it another time.

5. Go to **Settings**.

6. Scroll down or type **General** on the search bar. Tap **General**.

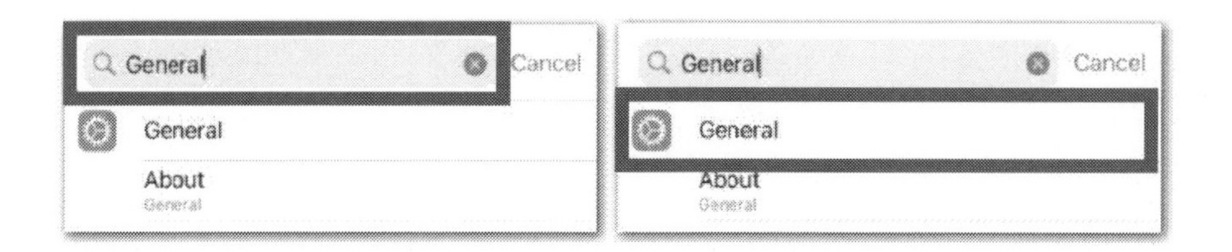

7. Tap **Software Update**. The iOS version that is currently installed is displayed on the screen as well as when an update is available. If an update is available, tap **Download and Install**.

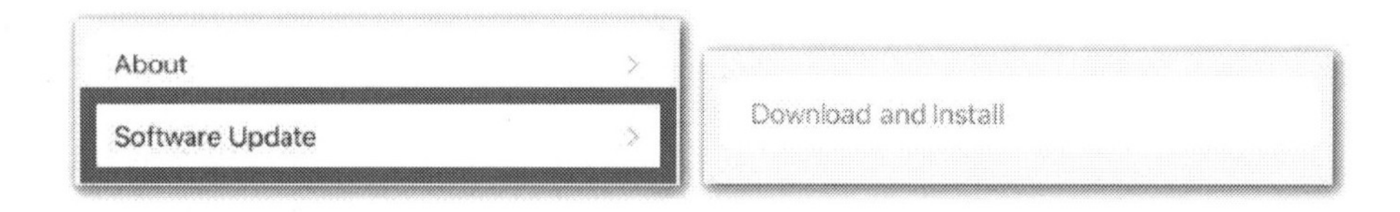

After the iOS update, try to see if the app is functional again.

Step 4: Delete and reinstall app

1. From the home screen, push and hold the crashing app icon, then tap **Remove App**.

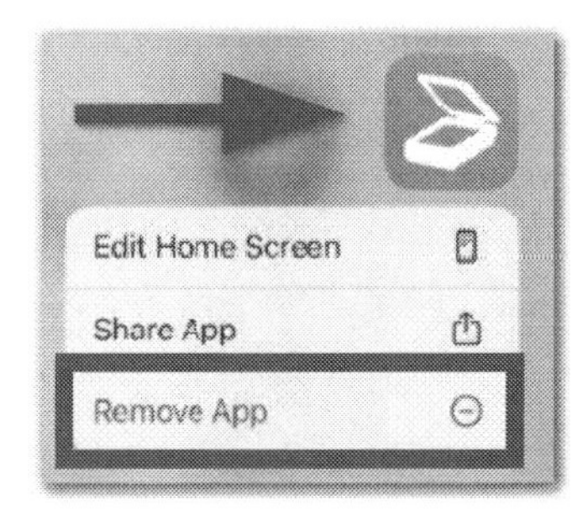

2. Go to **App Store**.

3. Go to **Search**, then type the name of the crashing app on the search bar.

4. Tap the app name from the search result, then tap the **download icon**. After the installation, try using the app again.

What If My iPhone is Overheating

When you first set up your smartphone, restore from a backup, wirelessly charge your device, use processor- or graphics-intensive apps, games, or features, such as augmented reality apps, or stream high-definition video, your device can become warm.

These circumstances are typical, and after the procedure is finished, or when you are finished with your activity, your gadget will regain its usual temperature. You can continue using your device if it doesn't display a temperature warning. If your device gets too warm, chances are certain conditions trigger your phone to change performance or temperature, especially when it is left on in a hot environment or used for an extended period. Your phone may turn off the display and alert you with a temperature warning screen

Turning it off and relocating it to a cooler location (out of direct sunshine) will help it cool off rapidly, so you can start using it again. To turn off your device, push and hold the **side button**.

Most problems encountered in iPhone can be resolved through troubleshooting; however, if you have trie all the steps but your device still doesn't work properly, you may contact Apple Support for assistance c bring your device to the nearest Authorized Apple Service Center.

CONCLUSION

Apple's iPhone is known for its basic, user-friendly design and is perfectly recommended for seniors. However, most are worried about what other people think about how they learn and use technology, which therefore affects their learning and limits their capacity to experience this state-of-the-art phone. To address this matter, this book that helps seniors easily learn how to use an iPhone has been published.

Phone for Non-Tech-Savvy Seniors walks users through the process of setting up the device, sharing troubleshooting tips, and answering frequently asked questions in order to help users resolve the most common technical issues in the iPhone while also covering other important functionalities. These include topics about the navigation of basic functions, followed by one of the most efficient features and the Phone's virtual assistant: Siri. It also includes managing messages and calls using your carrier's service and communication over the internet. In addition, this book walks you through the powerful camera feature, media, gallery, browsing the web using Safari, and the main iPhone apps you need to know and incorporate to your day-to-day living. What is more, your device's privacy and security are covered, so you can learn ow to effectively protect your data with your iPhone. Lastly, extra services provided by Apple are added.

s promised, easy-to-understand and enjoyable iPhone tutorials were delivered. All the most important atures, techniques, and information are collected and packed in this guide to ensure its usefulness to senior users. Technical contents were translated into layman's terms to help you understand the guide better and encourage you to create exciting experiences with your iPhone.

he iPhone sets itself apart from other smartphones because of its quality, so don't be afraid to explore e capabilities of the best smartphone on the market. Read, learn, and apply—or, much better, read, learn, make mistakes, start over again, and then apply.

Made in the USA
Columbia, SC
11 August 2023